HANDBOOK OF CELLULAR CHEMISTRY

Annabelle Cohen
The College of Staten Island (CUNY)

FOURTH EDITION

ISBN 0-7872-0484-6

Printed in the United States of America.
10 9 8 7 6 5 4 3 2 1

To Roslyn *in memoriam*

PREFACE

This text was originally written to meet the needs of students in allied health and other biology-based curricula who have little or no background in chemistry. My aim was, and is, to give the reader a reasonable understanding and appreciation of what lies at the heart of all biological sciences, namely, the structural and functional chemistry of living systems.

In the years that have passed since I wrote the first edition of the ***Handbook of Cellular Chemistry,*** there have been many, sometimes momentous, advances in our knowledge of cell membrane dynamics, functional proteins, genes, metabolic mechanisms, physiological messenger molecules, and viruses, to mention but a few. However, this edition, like previous editions, is intended to be a concise introductory guide rather than a comprehensive treatment of this complex and rapidly expanding field of science. The format and much of the material of earlier editions have been retained, but I have re-written, rearranged, and extensively updated various sections of the text. I have also stressed important new discoveries and concepts wherever relevant. This edition also includes chapter by chapter self tests at the back of the book, each of which features key terms to be learned, and study questions relating to the text material presented in the given chapter. It is my hope that students will be sufficiently interested and motivated by the information they find in this small book to pursue the study of the molecular basis of life in greater depth.

I am indebted to my son, David, for his invaluable help in processing the manuscript for this edition, and to both my sons for many hours of intellectually stimulating dialogue. It is also a pleasure to thank my friend and colleague, Dr. Jacqueline LeBlanc, Chairperson of the Biology Department, who has been a continuing source of encouragement and support.

Annabelle Cohen

CONTENTS

1

INTRODUCTORY CHEMISTRY FOR BIOLOGISTS

If a cell or a tissue is removed from the body and analyzed in a test tube, it will be found to consist of a mixture of certain chemical substances, such as water, salts, sugars, proteins, etc. These are the raw materials from which living matter is made. Nonliving matter, for example a grain of sand or a pebble, is similarly composed of distinctive chemical substances. *Matter*, living and nonliving, makes up the fabric of the universe. It is, by definition, anything that occupies space and has mass or weight. The basic chemical units of matter are *atoms*. In turn, groups of atoms combine to form more complex units called *molecules*. The study of the atoms and molecules of matter, their properties (characteristics), and the changes they undergo is the science of *chemistry*. Some knowledge of this science is essential for understanding the structure and function of living organisms.

matter

atoms

molecules

chemistry

ELEMENTS

The enormous numbers of substances found on earth, in both living and nonliving forms, are all made up of elements or combinations of elements. The term *element* is given to elementary chemical substances that cannot be further decomposed (broken down) into simpler substances by ordinary chemical means. Each element has its own specific type of atoms.

element

Some elements, such as gold, silver, copper, and sulfur, were known to the ancient world. As man's knowledge of chemistry advanced, more of the earth's elements were identified until what appeared to be a reasonably complete list of 92 naturally occurring elementary substances was assembled. These ranged from the lightest known element, hydrogen, to the heaviest, uranium. At present, there are considered to be 109 elements; the additional 17 elements have been made by man in various accelerator devices (such as cyclotrons).

Human tissues contain only about 20 of these elements. In fact, four of them, namely *carbon, hydrogen, oxygen,* and *nitrogen,* compose approximately 95% of the substance of the body (Fig. 1-1).

In chemical shorthand, each element is represented by a symbol. The symbol is the capitalized initial letter of the name of the element, often with a second (or more rarely, third) small letter taken from the name. Symbols may be derived from the English name of the element, or from the element's name in some other language, most usually Latin. Several examples of chemical symbols based mainly on the Latin names of elements are listed in Table 1-1.

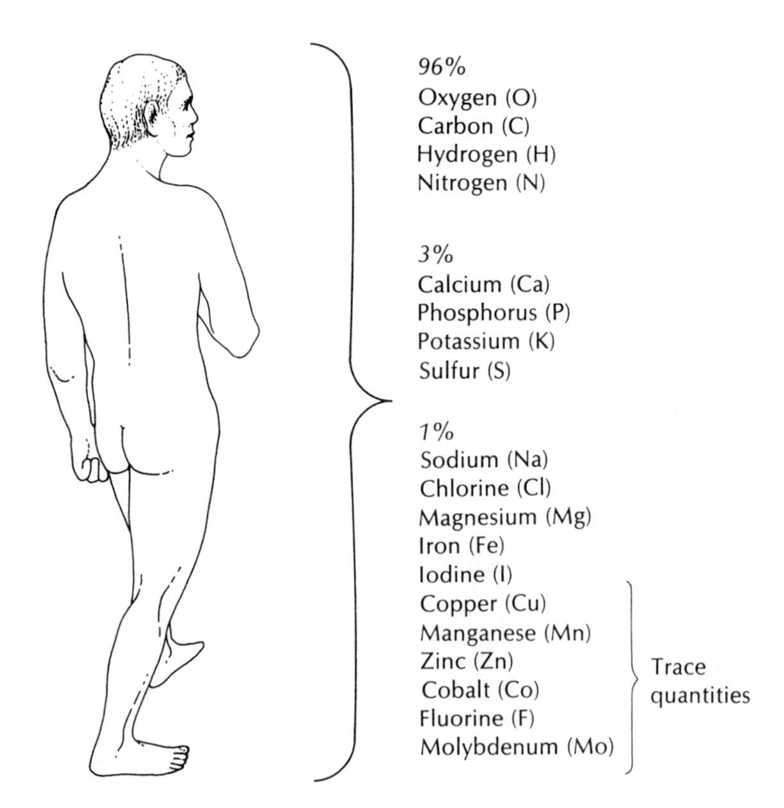

Fig. 1-1. The elements present in human tissues (approximate percentage by weight).

Table 1-1. Chemical symbols based on Latin names of elements

Element	Chemical symbol	Latin name
Antimony	Sb	Stibium
Copper	Cu	Cuprum
Gold	Au	Aurum
Iron	Fe	Ferrum
Lead	Pb	Plumbum
Mercury	Hg	Hydrargyrum*
Potassium	K	Kalium
Silver	Ag	Argentum
Sodium	Na	Natrium
Tin	Sn	Stannum
Tungsten	W	Wolfram†

* Literally meaning "liquid silver"; mercury is a liquid metallic element.
† Wolfram is the German name for tungsten.

Atomic Structure of Elements

Elements are composed of extremely small, discrete particles called atoms (the mass of an atom is in the range of 10^{-24} to 10^{-23} g; see appendix to this chapter for an explanation of the magnitude of these numbers.) The *atom* is the smallest unit of an element that exhibits all the characteristic properties and undergoes the characteristic chemical changes of that element. Atoms of the same element are more or less alike and differ in their properties from atoms of other elements. It is important to remember that the symbol for an element not only represents the *name* of the element, but is also the chemist's way of representing *one atom* of that element.

atom

For many years it was considered that atoms were indivisible units that could not be broken down into anything smaller. However, it is now known that atoms are clusters of varying numbers of even smaller (subatomic) particles called *protons, electrons,* and *neutrons.** The arrangement of these particles in an atom resembles in some respects a solar system of planets orbiting around a sun. The mass of an atom is concentrated in a positively charged *nucleus* made up of all the protons and neutrons of that particular atom, whereas the electrons orbit the nucleus at relatively great distances and in various energy levels, or shells.

nucleus

The subatomic particles are classified as follows:

* There are other subatomic particles, but they are not relevant to this discussion.

Table 1-2. Fundamental atomic particles

Name	Symbol	Location in atom	Assigned mass (d or amu)	Charge
Electron	e or e−	Outside of nucleus	0	−1
Proton	p	Nucleus	1	+1
Neutron	n	Nucleus	1	0

protons

electrons

neutrons

1. *Protons* are positively charged particles (having a +1 charge) with an assigned mass of 1 atomic mass unit. They are found in the nucleus of every atom.
2. *Electrons* are negatively charged particles (having a −1 charge) with a mass of about 1/1,840 of an atomic mass unit (protons are about 1,840 times heavier than electrons). They are found moving around outside the nucleus at various distances from it.
3. *Neutrons* are uncharged (neutral) particles with a mass approximately equal to the mass of a proton, that is, 1 atomic mass unit. They are found in the nucleus of every atom except the common hydrogen atom.

atomic mass unit
dalton

Note that the mass of either a proton or a neutron is considered to equal approximately *one atomic mass unit* (amu). Chemists also frequently express atomic mass in terms of the *dalton* unit *(d);* that is, 1 d equals 1 amu. The dalton (or atomic mass unit) is an arbitrary unit defined as $\frac{1}{12}$ the mass of the most common naturally occurring form of carbon, namely carbon 12 (^{12}C). This carbon atom has 6 protons (and 6 electrons) and 6 neutrons and has been assigned the standard atomic mass of exactly 12.000 d. Because the mass of an electron is negligible when compared with that of either a proton or a neutron, it is generally considered to have no mass. The characteristics of these subatomic particles are summarized in Table 1-2.

Electron Shells

electron shells

The atom, as described by the Danish physicist Niels Bohr in the early years of the twentieth century, was visualized as consisting of a very small, dense, positively charged nucleus, with electrons moving in circular or elliptical orbits at fixed distances around the nucleus. According to this model, an atom can have as many as 7 *electron shells*. A shell is defined as a discrete volume around a nucleus in which a given electron or set of electrons moves. Each shell can hold only a given maximum number of electrons. The first shell (nearest the nucleus) can hold a maximum of 2 electrons; the second, a maximum of 8; the third, up to 18; the fourth and fifth, a maximum of 32; and so on. Shells are often referred to as *energy levels* and are identified by the letters K (the first shell) through Q for the successive shells up to the seventh.

energy levels

Regardless of the total number of electron shells, the outermost shell never contains more than 8 electrons; if the atom has just 1 shell (as is found only in the two lightest elements, hydrogen and helium), then it can be completely filled by only 2 electrons.

Diagrams of several representative atoms are shown in Fig. 1-2. The number of neutrons (n) and protons (p) making up the nucleus of each atom are indicated in the central circle. The electrons (e) are shown as numbers of e at the appropriate energy levels outside the nucleus. The shells, or energy levels, are numbered at the right side of the atoms from the nucleus outward.

Note that hydrogen has only 1 electron in only 1 shell; thus its outermost shell is incompletely filled. Helium has 2 electrons in its only shell, which is therefore completely filled with the maximum number of electrons (2) for that shell. Carbon has 2 shells, the first of which is filled with the maximum 2 electrons; the second, the outermost, is incompletely filled with 4 electrons. The sodium atom has 11 electrons in 3 shells. The first is filled with 2 electrons; the second, with 8; and the third shell, the outermost, is incomplete with only 1 electron. Chlorine has 17 electrons, likewise distributed in its 3 shells, with 2 electrons in the first, 8 in the second, and 7 in the incomplete outer third shell. The electrons in an incompletely filled, or unsatisfied, outermost shell are extremely significant because these are the *valence electrons* that make an atom chemically reactive (see Chapter 2).

valence electrons

It must be mentioned at this point that the Bohr atom model cannot account for certain aspects of the physical behavior of electrons in atoms. The modern theory of atomic structure is based on *quantum mechanics*. Although the quantum mechanical model of the atom is similar to the Bohr model, it embodies a considerably more complicated three-dimensional spatial arrangement of electrons in energy levels and sublevels. The quantum model of electron configuration takes into account the fact that electrons are in rapid, random motion and therefore exhibit wave-like properties in addition to particle properties. Because of their dual nature, it is impossible to place them in fixed positions on two-dimensional spherical orbits around the nucleus. Rather, electrons are considered to occupy a rather complex series of

quantum mechanics

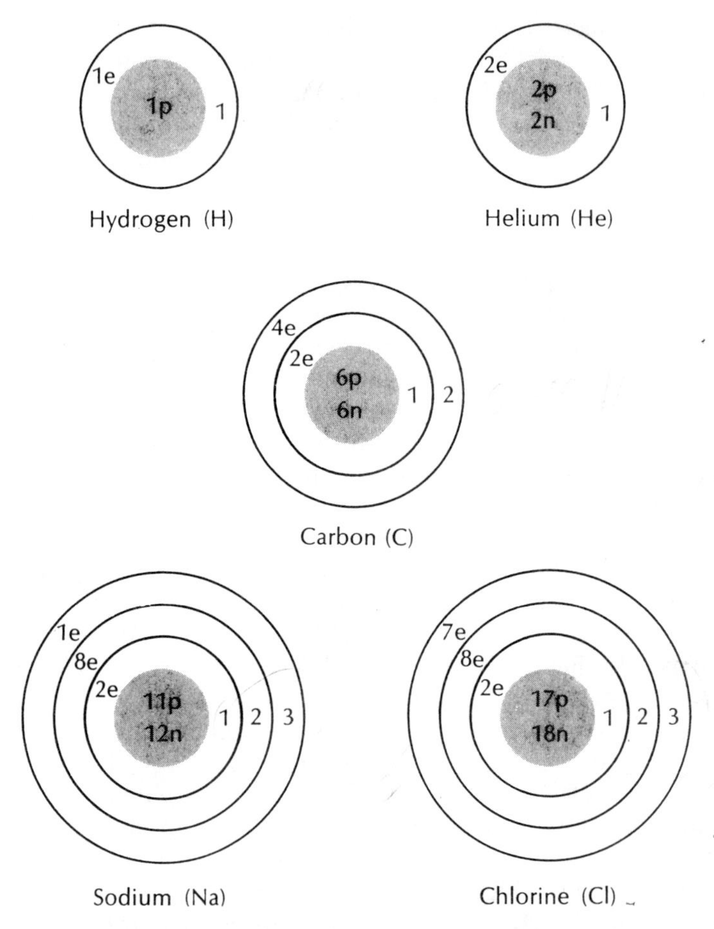

Fig. 1-2. Atomic structures of selected elements, indicating numbers and locations of protons (p), electrons (e), and neutrons (n). These drawings represent cross sections through the nuclei and three-dimensional, spherical electron shells. The sizes of the nuclei are greatly exaggerated; on this scale, each nucleus would actually be much smaller than a pencil point.

orbitals

principal quantum numbers *(n)*

orbitals of varying sizes, shapes, and energies. The orbital is defined as a region in the space around the nucleus where any particular electron has a 90% probability of being found at any given moment. The *principal quantum numbers (n)* 1, 2, 3, 4, and so on, correspond to the successive shells K, L, M, N, and so on, and relate generally to the average distance of the electrons from the nucleus; that is, electrons with the same principal quantum number are estimated to be about the same distance from the nucleus of the atom. The term *energy level* (which, as noted previously, is often used in the same context as *shell*) refers to the fact that electrons in the orbitals closest to the nucleus have the lowest energy, whereas those in orbitals further away from the nucleus (with increasing *n* values) have increasing energies.

The details of the modern version of electron configuration based on quantum theory are outside the scope of this text. However, students should visualize electrons, protons, and neutrons, not in terms of tiny static bodies fixed in time and space but as swift-moving packets of flashing energy. The subatomic particles are, in fact, entities on the borderline between energy and matter in our universe. An important concept to keep in mind is that the energy of the atom *(atomic energy)* is the source of the energy given off by the sun *(solar energy)*. As we shall see later, the sun is the primary source of the energy used by all living organisms on this planet.

atomic energy
solar energy

Atomic Number

The atoms of a given element contain a specific number of *protons* in their nuclei. This proton count is the *atomic number* of that element. There are at present 109 known elements. The range of atomic numbers is from 1 to 109; for example, atoms of the lightest element, hydrogen (H), contain 1 proton; atoms of the most recent man-made element (the heaviest, so far), contain 109 protons.

atomic number

In a neutral atom (an atom in an uncombined state) the number of protons inside the nucleus (the atomic number) *is the same* as the number of electrons outside the nucleus; that is, the number of positive charges is balanced by the number of negative charges. The importance of this fact is that the basic physical and chemical properties of an atom depend on the number of protons in its nucleus and the number of orbiting electrons. The atomic number is therefore the most fundamental property of an atom. If the atomic number changes for some reason (as it does in certain radioactive elements), the properties of that atom change quite drastically; in fact, the element undergoes a transformation into another element.

Mass Number

The *mass number* of an atom is the total number of particles in the *nucleus* (protons plus neutrons). It is called the mass number because each proton and each neutron constitutes one unit of atomic mass, and the sum of the units accounts for most of the

mass number

mass of an atom; the mass of electrons in an atom is negligible. (One atomic mass unit, or dalton, is defined as $\frac{1}{12}$ the mass of an atom of carbon 12.)

The complete chemical notation for an element gives its mass number and atomic number:

$$^{A}_{Z}E$$

where

A = mass number (in daltons)
E = symbol of the element
Z = atomic number of the element

Examples of this type of notation are:

1. Most atoms of hydrogen (H) have 1 proton and no neutrons in the nucleus. Therefore, the atomic number of H is 1; the mass number of H is 1 d. An atom of this element is thus represented as

$$^{1}_{1}H$$

2. Most atoms of helium (He) have 2 protons and 2 neutrons in the nucleus. The atomic number of He is 2; the mass number of He is 4 d. An atom of this element can be shown as

$$^{4}_{2}He$$

3. Most atoms of chlorine (Cl) have 17 protons and 18 neutrons. The atomic number of Cl is 17; the mass number of Cl is 35 d. Thus,

$$^{35}_{17}Cl$$

It will also be obvious from the above that an element's mass number (A) minus its atomic number (Z) equals the number of neutrons in the atoms of that element.

Isotopes

When the mass numbers of the atoms of a given element are compared, it is found that all of the atoms do not have the same mass number. For example, although the nuclei of 99% of all hydrogen atoms consist of 1 proton only, a small percentage will be found to have 1 or 2 neutrons as well. Atoms of the same element having the *same* atomic number but *different* mass numbers (due to different numbers of neutrons in the nucleus), are called *isotopes* of that element. Generally speaking, the isotopes of a given element have the same chemical properties. This factor is controlled by the numbers of protons and electrons, which are constant for all isotopes of that element.

isotopes

All elements found in nature have one or more isotopes, but a great many "artificial" isotopes have also been made by man. As a result, although there are just over 100 different known elements, there are more than 1,400 different varieties of atoms. Some examples of isotopes of elements are shown in Fig. 1-3.

Atomic Weight

When classifying elements by their mass numbers, it is inconvenient to include the different mass numbers of all the isotopes of that element. Chemist usually refer to the mass numbers of the atoms of an element in terms of the *atomic weight* of the element. This is not really a weight but an *average* of the mass numbers of the isotopes. The standard for atomic weights is the mass number of the most common isotope of carbon (mass number, 12). Carbon 12 has been assigned an atomic weight of exactly 12.000 d. To say that a given atom has an atomic weight of 36 d. indicates that it is three times as heavy as an atom of carbon 12. Atomic weight is thus a relative weight. As a general rule, *the atomic weight of any element is approximately equal to the mass number (expressed in daltons) of its most common isotope* (the average of the mass numbers of the isotopes of an element is weighted in favor of the most abundantly occurring isotope). Thus:

atomic weight

1. Oxygen (O) has an assigned atomic weight of 15.999 d, indicating that its most common isotope is ^{16}O.
2. Chlorine (Cl), a mixture of 2 isotopes, about 75% ^{35}Cl and 25% ^{37}Cl, has an atomic weight of 35.453 d.
3. The atomic weight of carbon (C) is 12.011 d. Naturally occurring carbon is a mixture of about 99% carbon 12 and about 1% carbon 13.

Table 1-3 lists some important elements, their symbols, atomic numbers, and atomic weights.

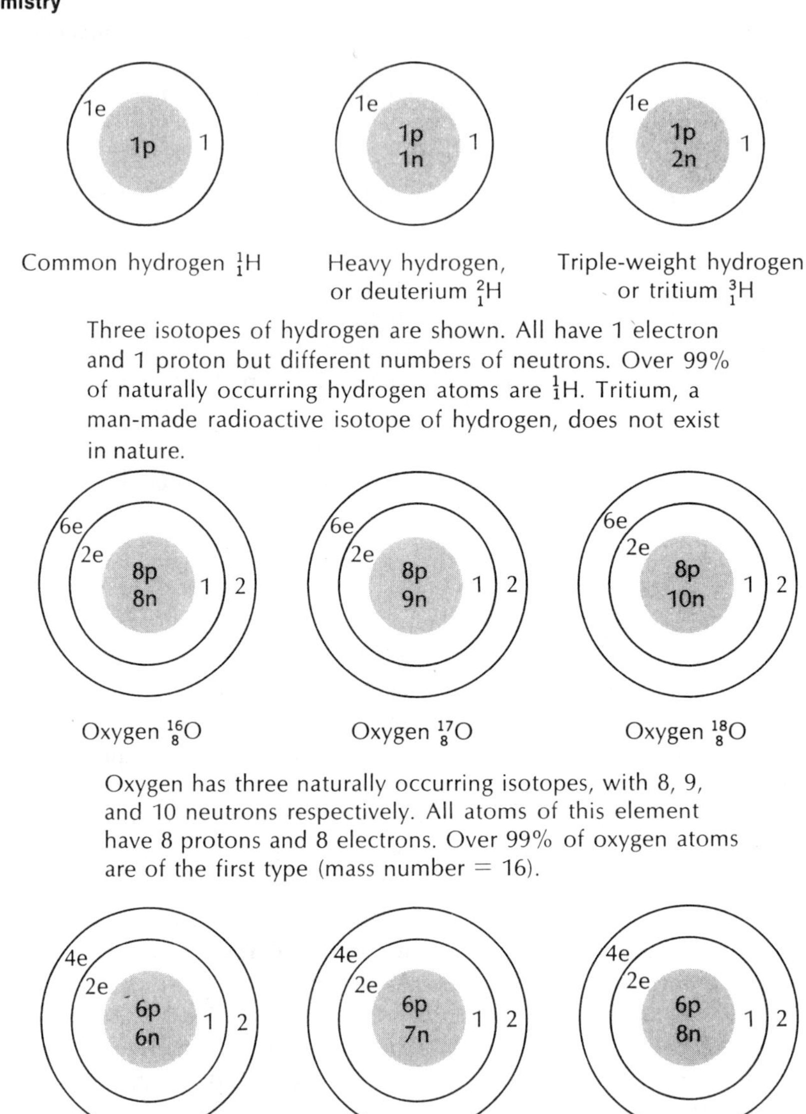

1e
1p
1
1e
1p
1n
1
1e
1p
2n
1
Common hydrogen ^1_1H
Heavy hydrogen, or deuterium ^2_1H
Triple-weight hydrogen or tritium ^3_1H
Three isotopes of hydrogen are shown. All have 1 electron and 1 proton but different numbers of neutrons. Over 99% of naturally occurring hydrogen atoms are ^1_1H. Tritium, a man-made radioactive isotope of hydrogen, does not exist in nature.
6e
2e
8p
8n
1
2
6e
2e
8p
9n
1
2
6e
2e
8p
10n
1
2
Oxygen $^{16}_8O$
Oxygen $^{17}_8O$
Oxygen $^{18}_8O$
Oxygen has three naturally occurring isotopes, with 8, 9, and 10 neutrons respectively. All atoms of this element have 8 protons and 8 electrons. Over 99% of oxygen atoms are of the first type (mass number = 16).
4e
2e
6p
6n
1
2
4e
2e
6p
7n
1
2
4e
2e
6p
8n
1
2

Common carbon $^{12}_6C$ Carbon $^{13}_6C$ Carbon $^{14}_6C$

Over 98% of naturally occurring carbon atoms are carbon 12. Carbon 14 is a radioactive isotope.

Fig. 1-3. Isotopes of common elements.

Table 1-3. Some important elements, their symbols, atomic numbers, and atomic weights

Element	Symbol	Atomic number	Atomic weight (d)*	Element	Symbol	Atomic number	Atomic weight (d)*
Aluminum	Al	13	26.9815	Lithium	Li	3	6.94
Antimony	Sb	51	121.7	Magnesium	Mg	12	24.305
Argon	Ar	18	39.94	Manganese	Mn	25	54.9380
Arsenic	As	33	74.9216	Mercury	Hg	80	200.5
Barium	Ba	56	137.3	Neon	Ne	10	20.17
Beryllium	Be	4	9.01218	Nickel	Ni	28	58.7
Bismuth	Bi	83	208.9806	Nitrogen	N	7	14.0067
Boron	B	5	10.81	Oxygen	O	8	15.999
Bromine	Br	35	79.904	Phosphorus	P	15	30.9738
Calcium	Ca	20	40.08	Platinum	Pt	78	195.0
Carbon	C	6	12.011	Potassium	K	19	39.10
Chlorine	Cl	17	35.453	Radium	Ra	88	226.0254
Chromium	Cr	24	51.996	Selenium	Se	34	78.9
Cobalt	Co	27	58.9332	Silicon	Si	14	28.08
Copper	Cu	29	63.54	Silver	Ag	47	107.868
Fluorine	F	9	18.9984	Sodium	Na	11	22.9898
Gold	Au	79	196.9665	Strontium	Sr	38	87.62
Helium	He	2	4.00260	Sulfur	S	16	32.06
Hydrogen	H	1	1.008	Tin	Sn	50	118.6
Iodine	I	53	126.9045	Tungsten	W	74	183.8
Iron	Fe	26	55.84	Uranium	U	92	238.029
Lead	Pb	82	207.2	Zinc	Zn	30	65.3

* Based on the assigned relative atomic mass of ^{12}C = 12.0000 d.

Radioactivity

The heavier isotopes of a number of elements are unstable; their nuclei decay spontaneously to more stable forms. This decay is called *radioactivity*. The earth is slightly radioactive from the presence of a few naturally *radioactive elements,* for example, radium (Ra), polonium (Po), and uranium (U), and from radioactive isotopes, or *radioisotopes,* of *stable elements,* for example, lead-204 (^{204}Pb) and potassium-40 (^{40}K). Additionally, about 1,000 radioisotopes have been made in nuclear reactors. The disintegration of a radioactive nucleus may be described as a tiny explosion in which one or more particles are ejected at high speeds. The particle emitted may be either an *alpha particle,* which consists of 2 neutrons and 2 protons, or a *beta particle* (actually an electron resulting from the disintegration of a neutron). In some instances, alpha and beta emission is accompanied by high-energy electromagnetic rays called *gamma rays*.

radioactivity

radioisotopes

alpha and beta radiation

It is interesting that the decay of a radioactive nucleus confirms the idea that the atomic number of an atom determines its properties and thus its identity. If a radioactive nucleus decays by emitting an alpha particle, a new nucleus is formed with a mass number that is four less than the original nucleus and an atomic number that is two less. If a nucleus decays by the disintegration of a neutron into an electron (emitted as a beta particle) and a proton that stays behind, then a new nucleus is formed. The new nucleus has the same mass number as the original, but an atomic number that is greater by one. In either case, a nuclear *transformation* has taken place, and a *different* atom, that is, a *different* element, is formed. For example, it can be readily demonstrated that when radioactive phosphorus-32 (^{32}P) decays by emitting a beta particle, it is transformed into the element sulfur:

$$^{32}_{15}P \xrightarrow[\text{decay}]{\text{radioactive}} {}^{32}_{16}S + \text{energy as beta radiation}$$

half-life

A quantitative estimation of the rate of decay of a radioisotope is expressed by its *half-life,* which is defined as the time required for half of the atoms in a given sample to disintegrate. The half-lives of radioactive elements found in nature are usually very long. For example, the half-life of uranium-238 is 4.5 billion years, that of carbon-14 is 5,760 years, and that of radium-226 is 1,590 years. There is considerable variation in the decay rates of elements, and shorter half-lives are measured in days, minutes, and seconds. The half-life of a radioactive isotope depends entirely on the structure of its atoms, rather than on any conditions, such as changes in temperature or pressure, that might be imposed on it.

The radiation emitted by radioactive atoms can be detected by instruments such as Geiger counters or scintillation counters. The technique of autoradiography is based on the fact that radiation also exposes photographic film; this technique is routinely used to demonstrate radioactivity in the tissues of experimental animals.

radioactive tracers

Radioactive isotopes in minute quantities have proved invaluable in biological research when used as *tracers*. A chemical compound labeled, or tagged with a radioactive element is processed by the body in exactly the same way as one containing the normal, nonradioactive form of the element, but the radiation it emits enables it to be quite easily tracked through even complex metabolic pathways. Since carbon and hydrogen are major components of all living cells, our knowledge of the chemistry of life processes has been immeasurably advanced by the use of various physiological substances labeled with radioactive carbon-14 or with tritium (^{3}H). In addition, radioactive tracers have been used to elucidate functions of specialized cells in the body. For example, the uptake of radioactive iron-59 into newly synthesized hemoglobin helped chart the development of specific bone marrow cells that give rise to red blood cells. Likewise, radioactive iodine-131 is taken up by the thyroid gland because some of the hormones produced by this gland contain iodine; by this means, the function (or malfunction) of the thyroid can be studied.

Radiation is highly energetic, whether it consists of charged particles, such as alpha or beta rays, or electromagnetic rays, such as x-rays or gamma rays. The energy of radiation can activate water molecules in the cells of the body to produce unstable complexes called free radicals (this effect is called ionization and, for this reason, radiation is usually referred to as *ionizing radiation*). Free radicals can react with cell constituents, such as enzymes, or the DNA molecules of chromosomes, causing lethal or sublethal interference with normal cell function. Cell death following lethal radiation is an indirect effect of irreversible damage to the functional apparatus of the cell. In low dosage, radiation can produce genetic mutations (see Chapter 13) and malignant transformation of cells, although such effects may take years to become apparent. Thus, exposure even to small amounts of radiation can prove to be hazardous.

ionizing radiation

The lethal effects of high-dosage ionizing radiation on cells are utilized in medicine for the treatment of malignant tumors. X-rays have been conventionally used for this purpose, but their use has been largely superseded by newer teletherapy units containing the man-made radioactive isotope, cobalt-60 (^{60}Co), which emits high-energy gamma rays.

PERIODIC TABLE

When the elements are arranged in order of *increasing atomic numbers,* it can be seen that there is a repeating pattern, or *periodic* recurrence at regular intervals, of various groups of elements with similar physical and chemical properties. We know that, in a neutral atom, the number of protons in the nucleus (the atomic number) is equal to the number of electrons orbiting around the nucleus. We have also learned that the properties of an atom are directly related to its atomic number and corresponding electron configuration. Adding these facts together results in a statement known in chemistry as the *periodic law;* namely the properties of the elements depend on their atomic structure and vary with their atomic number in a periodic (systematically recurrent) manner. This is the basis for the *periodic table* (Fig. 1-4), in which the elements are listed in sequences of increasing atomic numbers in such a way that elements with similar electronic configurations

periodic table

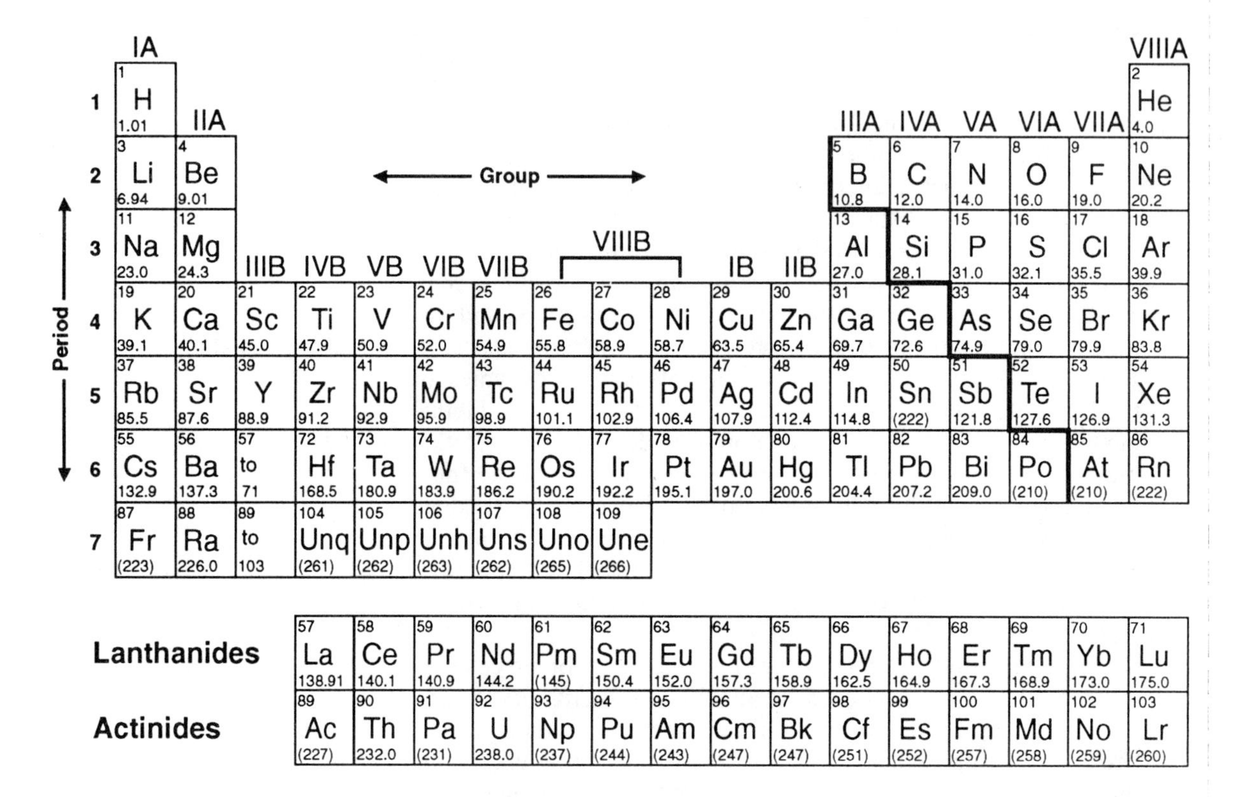

IA
IIA
IIIB IVB VB VIB VIIB
VIIIB
IB IIB
IIIA IVA VA VIA VIIA
VIIIA
Group
Period
1
2
3
4
5
6
7
1 H 1.01
2 He 4.0
3 Li 6.94
4 Be 9.01
5 B 10.8
6 C 12.0
7 N 14.0
8 O 16.0
9 F 19.0
10 Ne 20.2
11 Na 23.0
12 Mg 24.3
13 Al 27.0
14 Si 28.1
15 P 31.0
16 S 32.1
17 Cl 35.5
18 Ar 39.9
19 K 39.1
20 Ca 40.1
21 Sc 45.0
22 Ti 47.9
23 V 50.9
24 Cr 52.0
25 Mn 54.9
26 Fe 55.8
27 Co 58.9
28 Ni 58.7
29 Cu 63.5
30 Zn 65.4
31 Ga 69.7
32 Ge 72.6
33 As 74.9
34 Se 79.0
35 Br 79.9
36 Kr 83.8
37 Rb 85.5
38 Sr 87.6
39 Y 88.9
40 Zr 91.2
41 Nb 92.9
42 Mo 95.9
43 Tc 98.9
44 Ru 101.1
45 Rh 102.9
46 Pd 106.4
47 Ag 107.9
48 Cd 112.4
49 In 114.8
50 Sn (222)
51 Sb 121.8
52 Te 127.6
53 I 126.9
54 Xe 131.3
55 Cs 132.9
56 Ba 137.3
57 to 71
72 Hf 168.5
73 Ta 180.9
74 W 183.9
75 Re 186.2
76 Os 190.2
77 Ir 192.2
78 Pt 195.1
79 Au 197.0
80 Hg 200.6
81 Tl 204.4
82 Pb 207.2
83 Bi 209.0
84 Po (210)
85 At (210)
86 Rn (222)
87 Fr (223)
88 Ra 226.0
89 to 103
104 Unq (261)
105 Unp (262)
106 Unh (263)
107 Uns (262)
108 Uno (265)
109 Une (266)
Lanthanides
57 La 138.91
58 Ce 140.1
59 Pr 140.9
60 Nd 144.2
61 Pm (145)
62 Sm 150.4
63 Eu 152.0
64 Gd 157.3
65 Tb 158.9
66 Dy 162.5
67 Ho 164.9
68 Er 167.3
69 Tm 168.9
70 Yb 173.0
71 Lu 175.0
Actinides
89 Ac (227)
90 Th 232.0
91 Pa (231)
92 U 238.0
93 Np (237)
94 Pu (244)
95 Am (243)
96 Cm (247)
97 Bk (247)
98 Cf (251)
99 Es (252)
100 Fm (257)
101 Md (258)
102 No (259)
103 Lr (260)

Fig. 1-4. Periodic table of the elements. Atomic numbers are indicated above the symbols of the elements, atomic weights below. Numbers in parentheses represent the mass numbers of the most stable isotopes of radioactive elements.

(and similar properties) can be grouped together. Each element in the periodic table is shown in a box with the following typical notation:

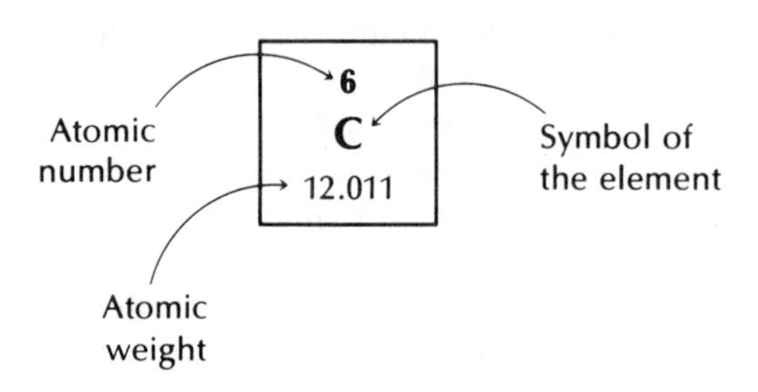

As shown in Fig. 1-4, the elements are arranged in the periodic table in horizontal rows, called *periods,* and in vertical columns, called *groups*.

periods

groups

In general, elements are classified as either *metals, nonmetals,* or *metalloids*. Most of the elements, 84 out of the current total of 109, are *metals*. These include all of the elements in Groups IA (with the exception of hydrogen, which is a gas), IIA, and Groups IB through VIIIB. There are also several metallic elements in Groups IIIA to VA, such as aluminum (Al), tin (Sn), and lead (Pb). Note that the two rows of elements at the bottom of the periodic table (the lanthanides and actinides) are actually metals in Group IIIB. Metals have certain distinguishing properties, that is, they are generally shiny, malleable (can be flattened out by hammering), ductile (can be drawn into wire), and are good conductors of heat and electricity. With a few exceptions, metals have 3 or less outermost electrons, and tend to combine chemically with nonmetals.

metals

Metallic properties tend to disappear as one proceeds from left to right in each period of the table. Thus, the *nonmetals,* which lack metallic properties, are all on the right side of the periodic table, starting with carbon (C) in Group IVA, with some of the elements in Groups VA and VIA, and all of the elements in Groups VIIA and VIIIA. Hydrogen (H), which is also a nonmetal, is the exception here once again. The change from metallic to nonmetallic properties is a gradual one. Occupying a roughly diagonal row from the upper right center to the lower right corner of the periodic table is a series of elements that separates the metals from the nonmetals. These elements, termed *metalloids,* have properties that are intermediate between metals and nonmetals. There are 7 metalloids, namely boron (B), silicon (Si), germanium (Ge), arsenic (As), antimony (Sb), tellurium (Te), and polonium (Po).

nonmetals

metalloids

The horizontal row or period that an element occupies is based on the capacity of the various energy levels for electrons, according to the quantum model of the atom. In simple terms, and for practical purposes, the period of an element can be said to correspond generally to the number of electron shells (or energy levels up to the known maximum of 7) present in the atoms of that element. Thus, the elements in period 1 have only 1 shell; the elements in period 2 have 2 shells; those in period 3 have 3 shells, and so on, up to period 7. Note that period 1 contains only 2 elements, hydrogen and helium. Periods 2 and 3 contain 8 elements each. Periods 4 and 5 are long, containing 18 elements each. Period 6 is a very long period, containing 32 elements. The last, period 7, is incomplete but theoretically would also contain 32 elements.

These numbers of elements constituting the periods, that is, integers of the order of 2, 8, 18, and 32, correspond to the maximum number of electrons a given, completely filled energy level can hold, starting with the shell closest to the atomic nucleus (the K shell, see p. 5). Since the mass of the atom also correlates with the atomic number, but in a more variable manner, there is a gradual progression of atomic weights in each period, from atoms with a lighter mass (left side) to atoms with a heavier mass (right side). Generally, properties of the elements in any particular periods change in a systematic pattern as one moves along the period.

The two lightest elements, hydrogen and helium, differ from all other elements in that their atoms have only the one K shell (outermost orbital), which, when completely filled, holds a maximum of 2 electrons. The helium shell is complete with 2 electrons, but hydrogen, with its half-filled outer orbital (1 electron), occupies a unique position among the other elements and, as we have seen, does not quite fit into any niche in the periodic table.

lanthanides
actinides

Period 6 and the incomplete period 7 are condensed in the table by removing the 14 *lanthanides* (lanthanum, $_{57}La$, to lutetium, $_{71}Lu$) and the 14 *actinides* (actinium, $_{89}Ac$, to lawrencium, $_{103}Lr$), and listing these elements separately in two rows at the bottom of the periodic table. As noted above, the 28 elements in these rows are Group IIIB metals; they are known as the *inner transition metals*.

inner transition metals

transition elements

Each vertical column or group of the periodic table consists of elements that have similar physical and chemical properties. The main-group elements, which have the more consistent periodic properties, are in the "A" groups; and the subgroup elements *(transition elements),* with more variable properties, are in the "B" groups at the center of the long periods. The common feature shared by all the elements in a group, or family, is that, regardless of the number of energy levels, or shells, in their atoms (this is indicated by the horizontal period the group elements occupy), they all generally have the same number of electrons in the outermost energy level. Recall that these outer electrons are the *valence* electrons that determine the chemical reactivity of the element. For example, it can be demonstrated that the elements of group IA, which recur at the *beginning* of each period, are all (with the exception of hydrogen) very reactive light metals. The group IA elements—lithium (Li), sodium (Na), potassium (K), rubidium (Rb), cesium (Cs), and francium (Fr)—are called the *alkali metals*. They have 1 outer valence electron and chemically combine with other elements to form similar classes of compounds.

alkali metals

On the other hand, if we observe the group VIIIA elements that recur at the *end* of each period (in the gray boxes at the extreme right of the table)—helium (He), neon (Ne), argon (Ar), krypton (Kr), xenon (Xe), and radon (Rn)—we find a family called the *inert gases* (or noble gases), all of which (with the exception of the lightest inert gas, helium) have outermost shells that are completely filled with the maximum number of 8 electrons. Helium, as noted previously, has 2 electrons that completely fill its single K shell. The inert gases as a group stand apart from all the other elements because they have little or no tendency to combine chemically. This correlation between the number of valence electrons and the chemical reactivity of the atom is discussed further in the next chapter. All that need be mentioned here is that the unreactive atoms of the inert gases have the most stable electron configuration, the *octet* (the maximum 8 electrons in the outermost shell, or, as in the case of helium, the maximum 2 electrons). Atoms with incompletely filled outer shells (and this applies generally to all elements other than the inert gases) attain the stable valence electron configuration of the inert gases by chemically combining with other atoms. Note that as one moves to the right horizontally in each period, the number of outer electrons in the atoms of each group progressively increases until it reaches the maximum filled capacity of the inert gas at the end of that particular period.

inert gases

Additional important families or groups of elements with similar properties are the elements of group IIA, the *alkaline earth metals,* all of which have 2 outermost electrons, and the elements of group VIIA—fluorine (F), chlorine (Cl), bromine (Br), iodine (I), and astatine (At)—a group of highly reactive, nonmetallic elements known as the *halogens* (salt producers), all of which contain 7 electrons in their outermost shell.

alkaline earth metals

halogens

It is an established fact that all elements with atomic numbers greater than 83 are radioactive. Element 92 (uranium, U) is the heaviest naturally occurring radioactive element. However, since 1940, scientists have created an additional 17 elements that do not exist in nature. The latter, from atomic number 93 to 109, are called the *transuranium elements* because they follow uranium in the periodic table. They were made by bombarding atomic nuclei with neutrons and other subatomic particles in cyclotrons and linear accelerators. One of them, plutonium ($_{94}Pu$), was in the nuclear bomb that destroyed Nagasaki. Another of these manmade elements, americium ($_{95}Am$), is now commonly used in household smoke detectors.

transuranium elements

The nuclei of the heavier synthetic transuranium elements at the end of the modern periodic table are very unstable. They tend to disintegrate so rapidly that most atoms disappear before they can be detected. For example, the one atom of element 109 that was reportedly created in 1982, survived for about 0.005 second before it blew apart. This is the reason why scientific claims in the area of the socalled 'superheavy' elements are often difficult to authenticate, and progress beyond element 109 has been very slow.

In spite of the complexities the periodic table presents, it has an appealing orderliness that is a reflection of the inherent symmetry and orderliness of atomic structure. The original periodic table was created in the last century by the Russian chemist Dmitri Mendeleev.* who based his classification of the then-known elements on atomic weights. Although there were discrepancies in his method, Mendeleev used the periodic law to make reliable predictions concerning the existence and properties of several elements that had not yet been discovered. Almost half a century later, the contributions of the English physicist H. G. Moseley finally led to the use of atomic numbers as the basis of classification of the elements; and the periodic table became, in its modern form, a systematized and accurate representation of fundamental chemical knowledge.

Appendix

Scientific Notation

Large and small numbers can be conveniently represented in a standard type of scientific notation in which the base 10 is raised to a *power,* or *exponent:* 10^x

Positive exponents represent multiplies of 10

10^0 = 1 (Note: Any number raised to zero power is equal to 1.)
10^1 = 10
10^2 = 100 (10×10)
10^3 = 1,000 ($10 \times 10 \times 10$)
10^4 = 10,000 ($10 \times 10 \times 10 \times 10$)

Negative exponents represent fractions of 10

10^{-1} = 0.1 = 1/10
10^{-2} = 0.01 = 1/100
10^{-3} = 0.001 = 1/1,000
10^{-4} = 0.0001 = 1/10,000

* Mendeleev is commemorated in the modern periodic table by the man-made element 101, mendelevium (Md).

Numbers expressed as the base 10 raised to an exponent conform to the following arithmetical rules:

1. To multiply, add the exponents algebraically.

   Examples
   $10^2 \times 10^3 = 10^5$
   $10^2 \times 10^{-3} = 10^{-1}$
   $10^{-2} \times 10^3 = 10^1$

2. To divide, subtract algebraically the exponent of the divisor from the exponent of the dividend (the number being divided).

   Examples
   $10^3/10^2 = 10^1$
   $10^{-3}/10^2 = 10^{-5}$
   $10^3/10^{-2} = 10^5$

3. The base 10 with a negative exponent can be changed to 10 with a positive exponent by expressing it as a reciprocal.

   Example
   $10^{-7} = 1/10^7$ (actually $10^0/10^7$)

4. The common logarithm of 10 raised to a power is equal to that power.

   Examples
   $\log 10^2 = 2$ $\quad$ $\log 10^{27} = 27$
   $\log 10^{-2} = -2$ $\quad$ $\log 10^{-27} = -27$

5. To work with other numbers (beside multiples of 10) in scientific notation, move the decimal point so that only *one* digit is on the *left* of the decimal point; then multiply this digit by 10 raised to an exponent corresponding to the number of places the decimal point was moved. If the decimal point was moved to the left, the exponent is positive; if it was moved to the right, the exponent is negative.

   Examples
   $17{,}800 = 1.7800. = 1.78 \times 10^4$

   $0.0000129 = 0.00001.29 = 1.29 \times 10^{-5}$

Metric System

Scientists throughout the world use the metric system, a decimal system, for measurements. In the metric system, the basic unit of length is the *meter* (m); the basic unit of mass is the *gram* (g); the basic unit of volume is the *liter* (L). Fractions and multiples of these units, all based on the number 10, are indicated by the following prefixes:

Prefix	*Symbol*	*Meaning*
Kilo	k	10^3 (one thousand) times the basic unit
Deci	d	10^{-1} (one tenth) times the basic unit
Centi	c	10^{-2} (one hundredth) times the basic unit
Milli	m	10^{-3} (one thousandth) times the basic unit
Micro	μ	10^{-6} (one millionth) times the basic unit
Nano	n	10^{-9} (one billionth) times the basic unit
Pico	p	10^{-12} (one trillionth) times the basic unit

Tables of metric units as applied to mass, length, and volume are shown below.

Mass

Gram (g) = basic unit	*Comparison with English weights and measures*
Kilogram (kg) = 10^3 g	1 lb = 453.6 g
Centigram (cg) = 10^{-2} g	1 kg = 2.205 lb
Milligram (mg) = 10^{-3} g	1 oz = 28.4 g
Microgram (μg) = 10^{-6} g	
Nanogram (ng) = 10^{-9} g	

Length

Meter (m) = basic unit	*Comparison with English weights and measures*
Kilometer (km) = 10^3 m	1 in = 2.54 cm
Centimeter (cm) = 10^{-2} m	1 km = 0.621 mi
Millimeter (mm) = 10^{-3} m	1 yd = 0.914 m
Micrometer (μm)* = 10^{-6} m	
Nanometer (nm) = 10^{-9} m	
Ångstrom (Å)† = 10^{-10} m = 0.1 nm	

Volume

Liter = basic unit	*Comparison with English weights and measures*
Kiloliter (kL) = 10^3 liters	1 qt = 0.946 liter
Centiliter (cL) = 10^{-2} liters	1 liter = 1.06 qt
Milliliter (mL) = 10^{-3} liters	1 fl oz = 29.57 mL

* The micrometer, often called a *micron (μ)* in biology, is used to measure cellular dimensions. For example, the diameter of a human red blood cell is 7 μm.

† The Ångstrom is an extremely small unit used to measure atomic and molecular dimensions. For example, the distance between the oxygen and hydrogen atoms of a water molecule is about 1 Å; the membranes of cells (composed of lipid and protein molecules) are about 80 to 100 Å thick.

Microliter (μL) = 10^{-6} liters
Nanoliter (nL) = 10^{-9} liters

Cubic centimeter (cc) = 1.00 cm^3 = 1.00 mL
Cubic millimeter (cmm) = 1.00 mm^3 = 1.00 μL

Sample Problem Using Scientific Notation and Metric System

The human red blood cell has a life span of 120 days, after which time it is removed from the circulation and broken down by phagocytic cells of the body. This means that 1/120 of the total number of red blood cells in the body, or approximately 200 billion red blood cells, is replaced each day.

Given

1. Normal average red blood cell count = 5 million cells/μL of blood
2. Normal total volume of blood = 5 liters

Calculation

1. 5×10^6 cells/μL blood,
   10^6 μL/liter

so,

$(5 \times 10^6) \times 10^6 = 5 \times 10^{12}$ cells/liter of blood

2. 5 liters blood/body

so,

$5 \times (5 \times 10^{12}) = 25 \times 10^{12} = 2.5 \times 10^{13}$ cells/5 liters blood/body

3. 1/120 of cells replaced per day
   $1/120 = 1/1.2 \times 10^2$

so,

$(1/1.2 \times 10^2) \times (2.5 \times 10^{13}) = (2.5/1.2) \times 10^{11}$ = approximately 2×10^{11} cells/day

Temperature

Temperature is recorded on thermometers and reported in degrees. The standard temperature unit used by scientists is based on the Centigrade, or Celsius, scale (C). The scale in common use in the United States is the Fahrenheit scale (F). On the Fahrenheit scale, the freezing point of water is 32°, and the boiling point of water is 212°. On the Celsius scale, the freezing point of water is 0°, and the boiling point is 100°. Human body temperature is 98.6°F, or 37°C.

To convert °C to °F: $F = \frac{9}{5}C + 32$
To convert °F to °C: $C = \frac{5}{9}(F - 32)$

Sample Temperature Conversion Problems

1. Convert 100°C (the boiling point of water) to °F:

$$\frac{9 \times 100}{5} + 32 = 180 + 32 = 212°F$$

2. Convert 98.6°F (normal body temperature) to °C:

$$\frac{5(98.6 - 32)}{9} = \frac{5 \times 66.6}{9} = 37°C$$

2

CHEMICAL BONDS AND COMPOUNDS

The isotopes of a given element have the same chemical properties. This factor is controlled by the numbers of protons & electons which are constant for all isotopes of that element.

All the atoms (isotopes) of a given element have the same electron configuration. This distribution of electrons confers on each element its chemical reactivity, that is, the type of combinations it will form with other elements. When the atoms of an element combine with other atoms in definite proportions by weight, a *chemical compound* is formed, and the atoms are said to be held together by *chemical bonds*.

chemical compound
chemical bonds

The formation of compounds involves chemical changes. The products of a chemical reaction often have very different properties from the reactants that combined to make them. For example, the gases hydrogen and oxygen in chemical combination form water; chemical combinations of carbon (as in coal), hydrogen (a gas), and oxygen (a gas) form products widely diverse from the original reactants, such as sucrose (or cane sugar).

Free, uncombined atoms are the exceptions in nature; it is far more common to find elements in the combined state. Literally millions of compounds exist, some of which are naturally occurring; many others are synthetically produced in the laboratory. The number is so great that it is customary to have specific branches of chemistry dealing with different classes of compounds. Thus, *organic* chemistry is the study of compounds containing the element carbon (organic compounds); *inorganic* chemistry deals with compounds that do not contain carbon (inorganic compounds): *biochemistry* is concerned with the structural and functional organic compounds found in living cells.

organic
inorganic
biochemistry

VALENCE ELECTRONS

valence electrons
octet rule

The electrons involved in the formation of chemical bonds are usually the electrons of the *outermost energy level* of an atom. These are called *valence electrons*. The key to the chemical reactivity of the atom is, in almost all cases, the *octet rule*, or the rule of eight. This rule is based on the observation that atoms interacting with other atoms tend to lose, gain, or share electrons to acquire a total of 8 (octet) electrons in their outermost energy level. The octet configuration is chemically the most stable; it is found in the inert gases of group VIIIA of the periodic table. These elements display very little or no chemical reactivity. A basic principle of chemistry is involved here; namely, when the outer orbitals of atoms are completely filled with the requisite number of electrons, they are, in a manner of speaking, reluctant to form chemical bonds. Conversely, when the outer orbitals of atoms are incompletely filled, they tend to compensate for the deficit by forming chemical bonds with other atoms. In doing so, they gain the stability of the inert gases.

electron dot symbol

A convenient representation of the valence electrons of an element is the *electron dot symbol*. The chemical symbol of the element indicates the nucleus and the inner shell electrons of the atoms; the outer energy level valence electrons are represented by dots placed around the symbol. For example, the electron dot symbol for hydrogen is

H·

Table 2-1 shows electron dot symbols for some of the more common elements. Electron dot formulas are also convenient repre-

Table 2-1. Electron dot symbols indicating valence electrons

H·													He:
Li·									·Ċ·	·N̈·	:Ö·	:F̤̈·	:N̤̈e:
Na·	Mg:							·Ȧl·	·Ṡi·	·P̈·	:S̈·	:C̤̈l·	:Ä̤r:
K·	Ca:	Mn:*	Fe:*	Co:*	Ni:*	Cu:*	Zn:				:S̈e·	:B̤̈r·	:K̤̈r:
	Sr:					Ag·			·Ṡn·*			:Ï̤·	:Ẍ̤e:
	Ba:					Au·*	Hg:*		·Ṗb·*				:R̤̈n:

The above arrangement of elements corresponds to their positions in the periodic table (Fig. 1-4).
The inert gases on the extreme right all have a stable configuration of valence electrons in the outer energy level (a duo in He, octets in the remainder).
* In some transition elements, an electron from an inner energy level may also act as a valence electron.

sentations of the chemical compounds formed by interacting atoms.

Atoms may acquire stable electron configurations by forming at least two distinct types of chemical bonds with other atoms:

ion

1. An atom may form an *ion* (charged species) by losing electrons out of or accepting electrons into its outermost shell. Oppositely charged ions may then form associations called *ionic bonds*.
2. Atoms may *share* electron pairs. The shared electrons are associated equally with both atoms forming the bond, thus filling the outermost energy levels of both atoms at the same time. Such bonds are *covalent bonds*.

IONIC BONDS

ionic bonds

In the formation of *ionic bonds,* one atom loses electrons and another atom gains them. An example of a compound formed by ionic bonds is the salt sodium chloride (NaCl), which consists of 1 atom of the element sodium and 1 atom of the element chlorine. By referring to the electron dot symbols of these elements, it can be seen that chlorine (with 7 valence electrons) could attain a stable octet configuration by either *losing* 7 electrons or by *gaining* 1. Likewise, sodium, (with 1 valence electron) could *gain* 7 or *lose* 1. The tendency is for the atom to lose or gain the *least* number of electrons necessary to acquire the octet configuration. Therefore, sodium gives up its 1 valence electron to the chlorine atom, filling chlorine's outer shell and eliminating its own partially filled outer shell. In the reaction, chlorine acquires the electron configuration of the inert gas argon; sodium, with its outer shell emptied and a filled inner shell exposed, acquires the stable configuration of the inert gas neon:

$$\underset{\text{atoms}}{\text{Na}\cdot + \cdot\ddot{\underset{\cdot\cdot}{\text{Cl}}}\!:} \rightarrow \underset{\text{ions}}{[\text{Na}^+\ \cdot][\,:\ddot{\underset{\cdot\cdot}{\text{Cl}}}\!:^-]} \quad \text{(NaCl, sodium chloride)}$$

Initially, both atoms were electrically neutral, containing equal numbers of protons and electrons. Now chlorine has *one more electron* than it has protons (17 protons, 18 electrons)—therefore it has one net *negative* charge; sodium has *one less electron* than protons (11 protons, 10 electrons)—therefore it has one net *positive* charge:

$$\underset{\text{neutral atom}}{Na^0} \rightarrow \underset{\text{ion}}{Na^+} + e^-$$

$$\underset{\text{neutral atom}}{Cl^0} + e^- \rightarrow \underset{\text{ion}}{Cl^-}$$

In the above notations, the superscript 0 indicates no charge (neutrality). The superscripts + and − indicate a net charge of one plus and one minus respectively.

cations
anions

A charged species (one with an unbalanced electrostatic charge), whether it is monoatomic (composed of 1 atom) or polyatomic (composed of more than 1 atom), is called an *ion*. Positively charged ions are *cations;* negatively charged ions are *anions*.* An ionic bond is simply an electrostatic association between two oppositely charged ions.

In the formation of sodium chloride, 1 atom of sodium and 1 atom of chlorine are involved. This is a proportion that depends on the number of valence electrons in the outer energy levels of the combining atoms. For example, if we consider formation of the salt calcium iodide, calcium has 2 valence electrons to give up, and an iodine atom requires only 1 electron to fill its outermost shell. Therefore, 2 iodine atoms combine with 1 calcium atom to form calcium iodide:

$$Ca{:} + 2\,{\cdot}\ddot{\underset{..}{I}}{:} \rightarrow [Ca^{2+}]\,2[I^-] \quad (CaI_2\text{, calcium iodide})$$

ratios of elements

The subscript 2 in the formula CaI_2 indicates that 2 iodine atoms are required for each calcium atom in this chemical combination. Other examples of different *ratios of elements* required to form compounds are as follows:

---

* The terms *cation* and *anion* refer to the fact that when two electrodes connected to a source of electric current are placed in a solution of ions, the negative electrode, or *cathode*, attracts the positively charged ions (cations) and the positive electrode, or *anode*, attracts the negatively charged ions (anions).

aluminum + chlorine → aluminum chloride ($AlCl_3$)

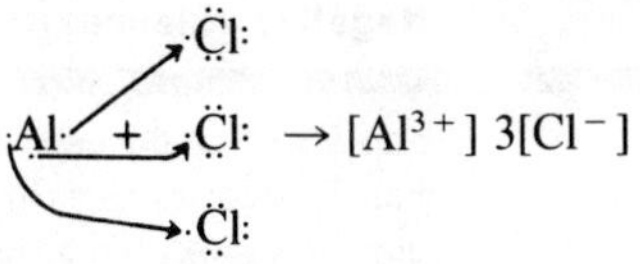

$$Al\cdot + 3\,\cdot Cl: \rightarrow [Al^{3+}]\,3[Cl^-]$$

potassium + sulfur → potassium sulfide (K_2S)

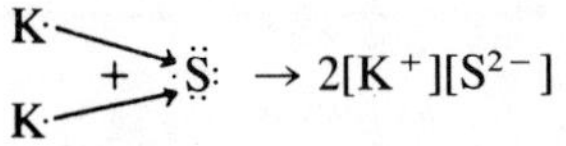

$$2\,K\cdot + \cdot S: \rightarrow 2[K^+][S^{2-}]$$

aluminum + oxygen → aluminum oxide (Al_2O_3)

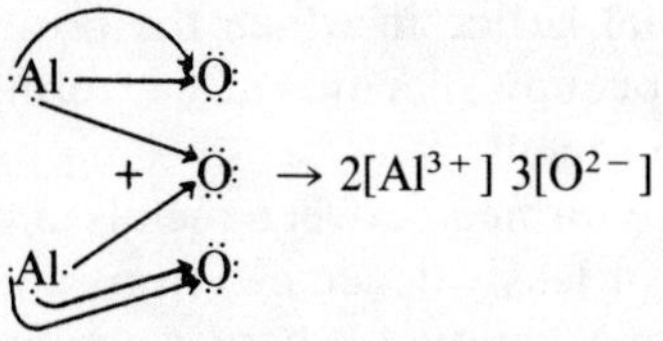

$$2\,Al\cdot + 3\,O: \rightarrow 2[Al^{3+}]\,3[O^{2-}]$$

The notations NaCl, CaI_2, $AlCl_3$, K_2S, and Al_2O_3 for the compounds shown above are chemical *formulas*. The formula expresses the elements present in the compound and the number of atoms of each element necessary to form the compound. The empirical formula also conventionally represents *one* unit (ionic aggregate or molecule) of the compound,* just as the symbol for an element represents one atom of that element. The number of atoms of each element in the compound is given by the subscript next to that element (note that when the subscript is 1, it is omitted). Subscript notation in the formula of any chemical compound is based on the *law of constant proportions: elements always combine chemically in fixed (definite) proportions by weight*. It follows that each compound has its own unique characteristic formula indicating the fixed ratio of the different types of atoms it contains.

formulas

law of constant proportions

Ionic bonds are not true chemical bonds. Rather, as noted above, they are electrostatic attractions between oppositely charged ions. (Opposite charges mutually attract and like charges mutually repel each other.) For general purposes, the electrostatic attraction between the elements in an ionic compound can be considered a type of chemical bond. However, if a solid *crystal*

crystal

* If *more* than one unit of a substance is to be indicated, the appropriate number must be written in *front* of the formula; for example, the formula NaCl represents one unit of sodium chloride; 2NaCl represents two units; 3NaCl represents three units, and so on. The significance of this notation will be seen at the end of this chapter when the balancing of chemical equations is discussed.

Table 2-2. Valences of some metals and nonmetals

Positive valence: metals	Negative valence: nonmetals
Group IA—alkali metals 1+ (tend to lose 1 electron)	Group VIIA—halogens 1− (tend to gain 1 electron)
Group IIA—alkaline earth metals 2+ (tend to lose 2 electrons)	Group VIA—oxygen group 2− (tend to gain 2 electrons)
Group IIIA—boron-aluminum group 3+ (tend to lose 3 electrons)	Group VA—nitrogen group 3− (tend to gain 3 electrons)

of an ionic compound is examined, the ions can be seen to exist as separate entities. For example, a crystal of sodium chloride (NaCl) is a three-dimensional lattice in which the positive Na^+ ions and negative Cl^- ions occupy specific sites according to the electrostatic forces acting on them.

In general, ionic bonds are formed between metals and nonmetals. Metals usually have 3 or less valence electrons and tend to lose them to form positive ions (cations), whereas nonmetals usually have 4 or more electrons and tend to gain others, thereby forming negative ions (anions). The general rule is that in the formation of ions, an element having less than 4 valence electrons will lose them and an element having more than 4 valence electrons will gain additional electrons to complete the outer shell.

valence

The term *valence* is used to indicate the number of electrons an element must lose or gain in order to acquire a stable configuration in its outermost shell. If electrons are to be lost, the valence has a plus sign; if electrons are to be gained, then the valence is negative. Table 2-2 is a list of periodic group elements showing positive and negative valences. The use of the more general term, *oxidation number*, rather than valence, is described on p. 33.

COVALENT BONDS

covalent bond

A chemical bond that results from the sharing of electron pairs between atoms is called a *covalent bond*. Unlike ionic bonds, which are formed between dissimilar atoms such as metals and nonmetals, covalent bonds can be formed between similar or even identical atoms. No actual transfer of electrons takes place. In most covalent bonds, each atom contributes 1 of the pair of shared electrons, and by this means each attains a stable octet (or, in the case of hydrogen, a stable duo) configuration. The chemical bond results from the overlapping of the two outer atomic orbitals of each atom. An atomic aggregate linked together by such bonds is called a *molecule*.

molecule

Molecules are readily formed by nonmetals. For example, atoms of the common gases—chlorine, hydrogen, oxygen, and nitrogen—exist naturally in the form of diatomic (2-atom) molecules. Chlorine and hydrogen molecules can be shown by the following electron dot formulas:

H : H   :C̤̈l : C̤̈l:

Note that each atom has its outer shell completed: each H has 2 electrons and each Cl has 8 electrons. The shared electrons are now revolving around both nuclei and are in such rapid motion that they simultaneously satisfy the needs of both atomic nuclei.

The above molecules can also be shown as structural formulas by representing each shared pair of electrons by a dash (and eliminating reference to the other electrons of the now filled outer shells)

H—H   Cl—Cl

and as molecular formulas, with the subscript indicating the number of atoms making up the molecule.

H_2 Cl_2

double bond

When 2 atoms of oxygen form a diatomic molecule, *two* pairs of electrons must be shared in order for each atom to attain a stable octet configuration. This type of bond is called a *double bond:*

Ö̤::Ö̤   O═O O_2

An example of a covalent compound with *two* double bonds is the gas carbon dioxide (CO_2). Carbon has 4 valence electrons, and oxygen has 6. They form a symmetrical molecule with the carbon atom in the center sharing two pairs of electrons with each of the oxygen atoms; the 3 atoms thus acquire a stable octet in the outer energy level of each:

Ö̤::C::Ö̤   O═C═O CO_2

triple bond

Nitrogen has 5 valence electrons. In the formation of a diatomic molecule of nitrogen, three pairs of electrons are shared between the 2 atoms. This is known as a *triple bond:*

$$\ddot{N}:::\ddot{N} \qquad N{\equiv}N \qquad N_2$$

The sharing of more than three pairs of electrons between two atoms does not occur.

POLYATOMIC IONS

polyatomic ions

Certain groups of covalently bonded elements may act together as a single ion in forming compounds. These aggregates of 2 or more covalently bonded atoms, which have an overall charge, are called *polyatomic ions* (or complex ions). Most of these ions have an excess of electrons when in chemical combination and are thus *anions*. They are often ionically bonded as a unit to metal cations. The one common positively charged *(cationic)* polyatomic ion is the ammonium ion (NH_4^+), which reacts chemically as a unit in a manner similar to univalent metallic cations, such as Na^+ and K^+. Polyatomic ions are common and play an important role in many cellular processes. Table 2-3 is a list of these ions with their symbols and valences.

Table 2-3. Polyatomic ions

Name	Formula	Valence*
Bicarbonate ion	HCO_3^-	1−
Carbonate ion	CO_3^{2-}	2−
Chlorate ion	ClO_3^-	1−
Permanganate ion	MnO_4^-	1−
Phosphate ion	PO_4^{3-}	3−
Monohydrogen phosphate ion	HPO_4^{2-}	2−
Dihydrogen phosphate ion	$H_2PO_4^-$	1−
Nitrate ion	NO_3^-	1−
Nitrite ion	NO_2^-	1−
Sulfate ion	SO_4^{2-}	2−
Sulfite ion	SO_3^{2-}	2−
Hydroxyl ion	OH^-	1−
Ammonium ion	NH_4^+	1+

* The valence (and the charge) is for the entire group.

POLAR COVALENT BONDS

Chemical bonds cannot all be classified as purely covalent or purely ionic in character. In principle, covalently bonded molecules would have no overall charge, because electron pairs would be equally shared and equally attracted to each atomic nucleus. However, a great many covalent bonds do not conform to this rule. They appear to have a partially ionic character; that is, regions (poles) of positive and negative charge exist within the molecule. In molecules of this type, it is evident that certain atoms are exerting a greater power of attraction for the shared electron pairs than others. This property of attracting the electrons in a covalent bond is called *electronegativity* and is exhibited strongly by nonmetallic elements such as fluorine (F), oxygen (O), nitrogen (N), and chlorine (Cl).

electronegativity

The greater the difference in electronegativity between any two elements forming a bond, the more unequally shared the electrons are likely to be. When a covalent bond has an uneven distribution of electrons, it is called a *polar* bond. If a molecule is appropriately shaped, a polar bond may result in a *polar molecule*. Molecules in which there is little or no separation of negative and positive charges are *nonpolar*. For example, the diatomic gas molecules discussed previously are symmetrical, relatively nonpolar molecules in which no specific part of the molecule is more negative or more positive than another. One may thus consider polar bonds to be an intermediate type of structure between the two extremes of ionic bonds and nonpolar covalent bonds.

polar

nonpolar

The molecular configurations of three common substances—water (H_2O), ammonia (NH_3), and methane (CH_4)—provide good examples of the importance of molecular shape in polar and nonpolar covalence. Water is a liquid at room temperature; ammonia and methane are gases. Each consists of a central atom bonded to a sufficient number of hydrogen atoms to satisfy the octet rule (duo rule in the case of the hydrogen atoms). Electron dot formulas for these substances are as follows:

```
   ..                        H
  :O:       H:N:H          H:C:H
  H  H        ..             ..
              H              H
```

In water and ammonia, the more strongly positive oxygen nucleus (8 protons) and nitrogen nucleus (7 protons) exert a greater electrostatic pull on the shared electron pairs than do the hydrogen nuclei, which consist of only 1 proton. As a result, the regions around the oxygen and nitrogen atoms are more negative than the

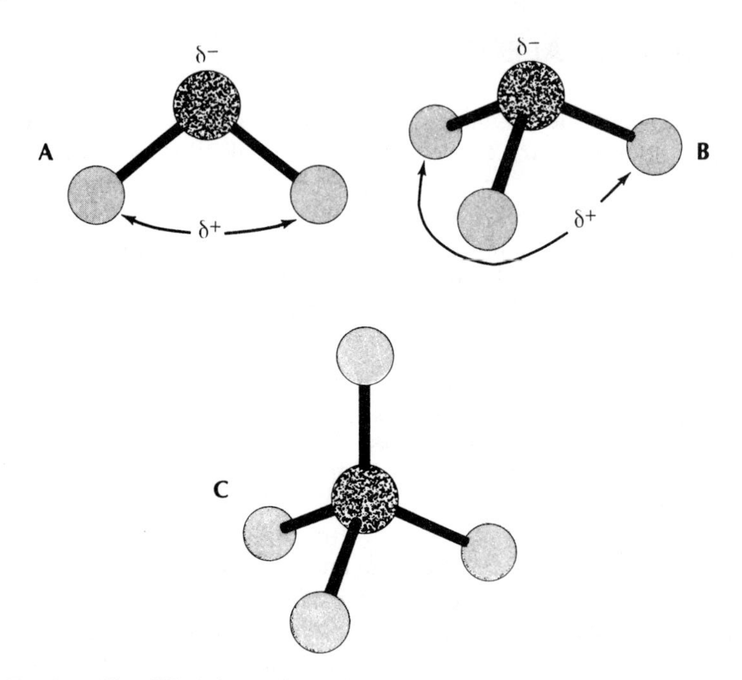

Fig. 2-1. Simplified three-dimensional structures of water. **A,** ammonia, **B,** and methane, **C.** δ+ and δ− indicate partial charges due to electronegativity of oxygen and nitrogen atoms. The smaller gray circles represent the hydrogen atoms in these compounds.

regions around the hydrogen atoms. Because of the angles of its bonds, the water molecule is not a linear molecule but, instead, has an angular shape. This shape, in conjunction with the charge separations of its polar bonds, gives the water molecule a partially negative end and two partially positive wings (Fig. 2-1, *A*). In three dimensions, the ammonia molecule is shaped like a pyramid, or a tripod. Each N—H bond is a leg, and all three legs are on the same side of the N atom. Therefore, the electronegativity of the N produces a polar molecule—ammonia has a negatively charged "head" end and three positively charged "feet" (Fig. 2-1, *B*).

The methane molecule (Fig. 2-1, *C*) is relatively nonpolar because of its *tetrahedral* (four-sided) shape (see Chapter 4 for more details on the three-dimensional aspects of carbon atoms). Since the H atoms are symmetrically distributed in space around the C atom, and since the C is only slightly more electronegative than the H, the covalent bonds in this molecule are considered to be predominantly nonpolar.

OXIDATION NUMBERS

oxidation
reduction

In chemistry, *oxidation* is defined as the loss of electrons by a substance and *reduction,* as the gain of electrons. When electrons are removed from elements that are being oxidized, they are gained by elements that are being reduced. The two processes thus take place simultaneously in what are known as oxidation-reduction reactions, or more briefly, redox reactions. (Biological oxidation-reduction reactions are described in Chapter 10.)

oxidation number

Although the common positive and negative valence numbers discussed previously are used to indicate losses and gains of electrons by substances forming relatively simple ionic compounds, they do not always provide an accurate picture of the electron exchanges that occur in complex oxidation-reduction reactions. For this purpose and also to extend the concept of valence to include all types of chemical combinations, a special kind of valence number, the *oxidation number,* is used. An oxidation number assigned to an element indicates the oxidation state of that element; that is, it represents the number of outermost electrons gained, lost, or unequally shared by any atom forming a chemical compound. It also indicates all the possible changes in oxidation state that an element may undergo, since elements can have several states of oxidation (and several oxidation numbers), depending on the number of electrons involved in the formation of the bonds in any particular chemical compound. Thus, although the usual oxidation number of hydrogen in most compounds is $+1$, it may also have an oxidation number of -1 in certain metallic hydrides, such as lithium hydride (LiH). In this type of compound, the hydrogen atoms are more electronegative than the metals and attract the electrons from the metal atoms.

Iron (Fe) and some other transition metals often contribute valence electrons from incomplete orbitals in their outermost shells, as well as in their next-to-outermost shells. Thus, iron forms ferrous and ferric types of compounds with oxidation numbers of $+2$ and $+3$, respectively. Certain elements, such as nitrogen (N), phosphorus (P), and sulfur (S), commonly form a variety of chemical combinations in which they may be assigned a fairly wide range of oxidation numbers. For example, oxidation numbers of nitrogen and phosphorus (each with 5 valence electrons) may range from -3 to $+5$, relative to the particular compounds they form. Sulfur, with 6 outermost electrons, has oxidation numbers from -2 to $+6$ in various chemical combinations.

Several of the more important rules for assigning oxidation numbers are as follows:

1. The oxidation number equals zero when the atom is in a neutral (uncombined) state (where all its electrons are present), or when 2 atoms of the same element equally share electrons in a diatomic molecule.
2. The oxidation number of a monoatomic ion equals its positive or negative charge (number of e^- removed or added).
3. Electrons shared unequally in polar covalent compounds are conventionally assigned to the more electronegative atoms. Thus, the oxidation number of oxygen in compounds such as H_2O and CO_2 is -2; the oxidation number of hydrogen in these compounds in $+1$ and that of carbon, $+4$.
4. The algebraic sum of the oxidation numbers of the atoms in the chemical formula of a compound is zero (total positive oxidation numbers equal total negative oxidation numbers).

A few simple calculations involving oxidation numbers and electron transfers in oxidation-reduction reactions are given at the end of this chapter.

HYDROGEN BONDS

Hydrogen, the lightest and smallest of the elements, possesses 1 electron orbiting in its 1 outer electron shell and 1 proton in its nucleus. In biological systems, hydrogen is very commonly found in covalent combination with highly electronegative atoms, such as in oxygen and nitrogen. Thus, the bonds formed between H and these elements are quite polar, possessing widely separated partial charges (Fig. 2-1). The H atoms carry partial positive charges because their electrons are closer to the O or N. Since the H nuclei (protons) lack significant electron shields, they are readily attracted to negative charges that may be in close proximity. The noncovalent associations resulting from such electrostatic forces are called *hydrogen bonds*.

hydrogen bonds

Hydrogen bonds form between the most electronegative atoms (in biological systems, these are invariably O or N atoms) and relatively *electropositive* hydrogen atoms that are already covalently linked to other electronegative atoms. In a small polar compound such as water, *intermolecular* hydrogen bonds may form between two or more molecules. In extremely large molecules (macromolecules such as proteins and nucleic acids), these bonds may be *intramolecular,* that is, between two or more parts of the same molecule.

By means of hydrogen bonds, polar molecules usually form fairly stable aggregates. Hydrogen bonds are a characteristic feature of water and are responsible for many of its unique properties (see Chapter 3 and Fig. 3-2). They are also important in stabilizing the three-dimensional structures of proteins and nucleic acids (see Chapters 7 and 11). The electrostatic attractions that result in hydrogen bonds are quite weak as compared to covalent bonds. As we shall see further on, the fact that hydrogen bonds can be easily broken and re-formed accounts to a large extent for the key role they play in molecular biology.

CHEMICAL SPECIES

Thus far, three identifiable chemical building blocks (species) of matter have been discussed:

1. *Atoms* are basic chemical species, the form in which elements exist. Each element has characteristic atoms. The two other chemical species are composed of atoms.
2. *Molecules* are made up of atoms that share electrons in covalent bonds. Some molecules are diatomic—made up of 2 combined atoms, for example, molecular oxygen (O_2). The large majority of molecules are polyatomic—made up of more than 2 combined atoms, for example, water (H_2O), glucose ($C_6H_{12}O_6$), and carbon dioxide (CO_2).
3. *Ions* are charged atoms, or charged groups of atoms. Ions are formed by the loss or gain of electrons. Cations (positively charged ions) are deficient in electrons, and anions (negatively charged ions) have an excess of electrons compared with the number of protons in the respective nuclei.

THE MOLE

The masses of the various chemical species—atoms, ions, and molecules—are all based on atomic weight. The *molecular weight* of a chemical compound is the sum of the atomic weights of all the atoms present in 1 molecule of the compound. If more than 1 atom of a particular element appears in the compound, as indicated by the subscript in the formula, its atomic weight must be multiplied by the subscript number. The term *formula weight* (rather than molecular weight) is often used for ionic compounds that are not considered molecules in the strict sense of the word,

molecular weight

formula weight

but the calculation is the same. Atomic, molecular, and formula weights are expressed in daltons (d) (or atomic mass units, amu). Chemists usually convert these units to grams for convenience of manipulation. The *gram atomic weight* (for atoms or monoatomic ions) or the *gram molecular weight* (for molecules or ionic compounds) is the atomic or the molecular weight of a species expressed in *grams*. One gram molecular weight (or gram atomic weight) of any chemical species is called a *mole* (or abbreviated, mol).

gram atomic weight
gram molecular weight

mole

Calculations

The ionic weight of a monoatomic ion is the same as the atomic weight of the neutral atom that formed the ion, since the mass of electrons is insignificant compared to the mass of atoms. Thus, 1 gram atomic weight of sodium (Na or Na^+) is 23 g; 1 mole of Na or Na^+ is 23 g. (The atomic weight of Na or Na^+ is 23 d.)

The molecular weight of water (H_2O) is 18 d, that is, 16 (O) plus 2 × 1 (2 H atoms). The gram molecular weight of water is 18 g. Thus, 1 mole of water weighs 18 g (18 g/mole H_2O).

One mole of the ionic compound sodium sulfate (Na_2SO_4) will equal its formula weight expressed in grams or, in general terms, its gram molecular weight. This is computed as follows:

	Atomic wt. of $2Na^+$ = 2 × 23	= 46
	Atomic wt. of 1S	= 32.1
	Atomic wt. of 4O = 4 × 16	= 64
	Total formula wt.	= 142.1 d
Formula wt. in grams	Gram molecular wt.	= 142.1 g
	1 mole of Na_2SO_4	= 142.1 g

Note that the formula weight of the sulfate ion (SO_4^{2-}) would be 96.1 d; 1 mole of the ion would equal 96.1 g.

It is important to consider at this point that the *formula* of a chemical species, as written, not only represents *one unit* (atom, ion, ionic compound, or molecule) of that species, but also *one mole*.

Avogadro's Number

Determining the weight of a mole of a particular species is not just an exercise in computation. The mole has physical significance in that *there are always the same number of molecules (or atoms or ions) in a mole of any chemical species*. This number is called *Avogadro's number (N)*, after the physicist who first established this relationship, and it is defined as the number of atoms contained in 12 g (1 mole) of carbon 12. The value of N is approximately 6.02×10^{23}. Thus, 1 mole of water (18 g), 1 mole of glucose (180 g), and 1 mole of DDT (354.4 g) all contain N number of molecules, specifically 6.02×10^{23}. The use of moles (or multiples or fractions of moles) is a great convenience in chemical calculations, since it quantitatively correlates the N *number* of units participating in a chemical reaction with the *weight* of that number.

Avogadro's number (N)

CHEMICAL EQUATIONS

Chemical shorthand employs symbols for elements, formulas for combinations of elements, and *chemical equations* to describe chemical reactions. A chemical equation for a hypothetical reaction may be recorded as follows:

chemical equations

$$A + B \rightarrow C + D$$

A and B, written on the left side, would be chemical formulas for one or more reacting substances *(reactants);* and C and D, on the right side, would be formulas for one or more substances produced by the reaction *(products)*. The arrow indicates the direction taken by the chemical reaction; that is, it is equivalent to stating that A reacting with B yields or produces C and D.

reactants

products

Many chemical reactions are *reversible*. The reaction will swing forward to the right and backward to the left until the rates of the opposing reactions are equal, at which point a *chemical equilibrium* is achieved. This is conventionally shown by arrows pointing in both directions:

reversible reaction

chemical equilibrium

$$A + B \rightleftharpoons C + D$$

Production of a gas during a chemical reaction is often indicated by an arrow pointing upward or by the notation (g), for gas, placed immediately after the formula. An insoluble product, or *precipitate*, may be shown by placing an arrow pointing downward or by the notation (s), for solid, after the formula of the substance. A substance dissolved in water (in aqueous solution) may likewise be indicated by the notation (aq). Examples of this type of notation in a chemical equation are as follows:

$$A(aq) + B \rightarrow C \downarrow \text{ or } (s) + D(g) \text{ or } \uparrow$$

balanced equations
law of conservation of mass-energy

To write an accurate chemical equation, the exact nature and correct chemical formulas of the reactants and products must be known and the equation must be *balanced.* Balancing an equation implies that the reaction as recorded must conform to the *law of conservation of mass-energy,* which states that matter (and energy) can neither be created nor destroyed. In practice, this means that every single atom (or ion) participating in the reaction must be strictly accounted for; that is, equal numbers of the same types of atoms must be present on both the reactant and product sides of the equation.

The most convenient way to balance chemical equations is to inspect them and employ a certain amount of trial-and-error computation. This procedure, as described below, is adequate for most equations. However, some oxidation-reduction reactions are more difficult to balance and require a count of electrons lost and gained to ensure that the net electrical charge on both sides of the equation is the same.

A simple (and explosive) chemical reaction that will serve to introduce the technique of balancing equations is the reaction in which hydrogen gas molecules combine with oxygen gas molecules to form water molecules:

$$H_2 + O_2 \rightarrow H_2O$$

A count of the atoms on both sides of this equation indicates that it is unbalanced; there are 2 hydrogen atoms and 2 oxygen atoms on the left side, and 2 hydrogen atoms and only 1 oxygen atom on the right side. As we have learned, the subscripts in the formulas are fixed properties of the reactant and product molecules and cannot be changed without changing the chemical nature of these substances. For example, attempting to balance the above equation by changing a subscript would result in

$$H_2 + O_2 \rightarrow H_2O_2$$

This would indeed balance the equation, but the statement would not be accurate. The reaction we are reporting specifies *water* as the product; the formula H_2O_2 does *not* represent water, but hydrogen peroxide, a powerful oxidizing agent used as an antiseptic and a bleach. The only valid way to balance the equation would be to place appropriate numbers (coefficients) *before* the formulas of the reactants and products. These numbers represent a quantity of units (atoms, ions, molecules, or *moles*) of each substance that is required for the reaction to take place in conformity with the principle of conservation of mass-energy. Coefficients are usually the smallest set of whole numbers that will suffice to balance the equation, but fractional coefficients are also sometimes used when convenient, for example, $\frac{1}{2}O_2$. By inserting the correct coefficients, we obtain a satisfactorily balanced equation for the reaction:

$$2H_2 + O_2 \rightarrow 2H_2O$$

Inspection of the numbers of each type of participating atom on both sides of the equation shows us that they are now equal; that is, there are 4 atoms of hydrogen and 2 atoms of oxygen on the left side, and 4 atoms of hydrogen and 2 atoms of oxygen on the right side. By adding these coefficients to the equation, we are in effect stating that 2 molecules (or moles) of hydrogen combine with 1 molecule (or mole) of oxygen to form 2 molecules (or moles) of water. Chemical reactions obviously involve astronomical numbers of molecules, but the balanced equation tells us that regardless of the quantities of molecules involved, the *ratio* of the quantities of hydrogen, oxygen, and water in this particular reaction is constant, that is, 2:1:2.

Rules for balancing equations can be summarized as follows:

1. Reactants and products must be known, and their correct formulas used.
2. Matter can neither be created nor destroyed; therefore, any given element participating in a reaction must be present on both the reactant and product sides of the equation; in addition, equal numbers of atoms of that element must appear on both sides of the equation.
3. In considering the notation in a balanced chemical equation, for example, in the notation $5X_2Y_4$, the formula subscripts 2 and 4 refer only to the combining proportions of the elements X and Y, respectively, whereas the coefficient 5 applies to the entire compound. Where no coefficient appears, such as in the notation X_2Y_4, a coefficient of 1 is implied.
4. For the purpose of computing the total number of atoms (or ions) of each element participating in a chemical reaction, *coefficient* $\times$ *subscript* = *total number of atoms (or ions).* Thus, the notation $5X_2Y_4$ in a balanced equation would indicate a total of 10 atoms (or ions) of element X and 20 atoms (or ions) of element Y.

Balancing Chemical Equations: Sample Problems

The simplest equations to balance are those that balance themselves; that is, where the ratios of participating substances are 1:1, as shown by the following examples:

1. Carbon (C) or sulfur (S) burn in the presence of oxygen (O_2) to form the gases carbon dioxide (CO_2) and sulfur dioxide (SO_2):

$$C + O_2 \rightarrow CO_2 \text{ (g)} \quad \text{(balanced)}$$
$$S + O_2 \rightarrow SO_2 \text{ (g)} \quad \text{(balanced)}$$

2. When carbon dioxide (CO_2) dissolves in water (H_2O), some of it chemically reacts to form carbonic acid (H_2CO_3):

$$CO_2 + H_2O \rightleftharpoons H_2CO_3 \quad \text{(balanced)}$$

   Note that under ordinary circumstances this reversible reaction is a rather slow one. In red blood cells, the reaction has great physiological importance and takes place very rapidly in the presence of the enzyme carbonic anhydrase.

3. The gas ammonia (NH_3) reacts with water (H_2O) to form the base ammonium hydroxide (NH_4OH):

$$NH_3 + H_2O \rightleftharpoons NH_4OH \quad \text{(balanced)}$$

   This is reversible reaction in which the ammonia gas, which has a characteristic pungent odor, is in equilibrium with ammonium hydroxide. The mixture is often used as a household cleaning agent.

Inspection of the above equations will show that there are equal numbers of atoms of each participating element on both sides of the equations. In the examples that follow, coefficients other than 1 have to be introduced in order to balance the equations.

4. Yeast cells ferment glucose ($C_6H_{12}O_6$) and produce ethyl alcohol (C_2H_5OH). In the reaction, the gas carbon dioxide (CO_2) is formed and given off:

$$C_6H_{12}O_6 \rightarrow C_2H_5OH + CO_2 \text{ (g)} \quad \text{(unbalanced)}$$

   The key element here is carbon; there are 6 C atoms on the left side and 3 C atoms on the right. Using a simple trial-and-error procedure, this can be remedied by inserting a coefficient of 2 before each product:

$$C_6H_{12}O_6 \rightarrow 2C_2H_5OH + 2CO_2 \text{ (g)} \quad \text{(balanced)}$$

   Note that there are now 6 C atoms on each side, and in the process, the H and O atoms are also automatically balanced.

5. Iron (Fe) rusts in moist air, combining with oxygen (O_2) to form the red compound, ferric oxide (Fe_2O_3):

$$Fe + O_2 \rightarrow Fe_2O_3 \quad \text{(unbalanced)}$$

The problem here is to balance 2 O atoms on the left side with 3 O atoms on the right. The simplest way to accomplish this would be to consider the lowest common multiple of these two numbers (6) and adjust the coefficients accordingly:

$$Fe + 3O_2 \rightarrow 2Fe_2O_3 \quad \text{(unbalanced)}$$

It is now necessary only to balance the Fe atoms:

$$4Fe + 3O_2 \rightarrow 2Fe_2O_3 \quad \text{(balanced)}$$

6. In the process of photosynthesis, plant cells utilize carbon dioxide (CO_2) and water (H_2O) in their environment to produce glucose ($C_6H_{12}O_6$) and oxygen (O_2):

$$CO_2 + H_2O \rightarrow C_6H_{12}O_6 + O_2 \quad \text{(unbalanced)}$$

On inspection, it can be seen that at least 6 molecules of CO_2 and 6 molecules of H_2O are required to generate the 6 C atoms and 12 H atoms that are present in 1 molecule of glucose:

$$6CO_2 + 6H_2O \rightarrow C_6H_{12}O_6 + O_2 \quad \text{(unbalanced)}$$

All that remains at this stage is to indicate a sufficient number of molecules of oxygen to account for the excess 12 O atoms on the left side of the equation:

$$6CO_2 + 6H_2O \rightarrow C_6H_{12}O_6 + 6O_2 \quad \text{(balanced)}$$

Oxidation Numbers and Oxidation-Reduction Reactions: Sample Problems

1. The following rules regarding oxidation numbers can be applied for calculating the oxidation number of any element in a given compound when the oxidation numbers of the other elements in the compound and the correct formula of the compound are known:

   Sum of oxidation numbers in formula = 0

   or,

   total positive oxidation numbers in formula = total negative oxidation numbers in formula

   and,

   oxidation number × subscript = total charge (total e^- lost, gained, or shared)

To calculate the oxidation number of sulfur (S) in the compound sulfuric acid (H_2SO_4), where the oxidation number of H is +1, and O is −2:

$$H_2^{+1}\ S^{x}\ O_4^{-2}$$

Total e^- of 2 atoms of H = 2 x (+1) = +2
Total e^- of 4 atoms of O = 4 x (−2) = −8

Therefore,

$$2 + x - 8 = 0$$
$$x - 6 = 0$$
$$x = +6 \text{ (oxidation number of S)}$$

Note that the charge on the entire sulfate (SO_4) group is −2. This may be derived from the algebraic sum of the oxidation numbers of O_4 (−8) and S (+6), or it may be inferred from the fact that since the total oxidation number of the 2 H atoms is +2, the entire sulfate group would have to be −2 for the algebraic sum of the oxidation numbers in the compound to equal zero.

2. In a balanced oxidation-reduction reaction:

   Net electrical charge (sum of oxidation numbers) of the reactants = net electrical charge (sum of oxidation numbers) of the products

   and,

   number of e^- lost = number of e^- gained

   This type of calculation can be demonstrated by referring to the balanced equation for ferric oxide (see example 5, p. 40), a simple oxidation-reduction reaction that is reproduced here with the various oxidation numbers of the two elements in parentheses above the formulas:

$$\overset{(0)}{4Fe} + \overset{(0)}{3O_2} \rightarrow \overset{(+3)(-2)}{2Fe_2O_3}$$

The *total electrical charge* on each atom in the equation will be calculated by *coefficient* × *subscript* × *oxidation number*. Since the oxidation numbers of the reactants Fe and O_2 are zero, the coefficients and subscripts are all multiplied by zero; thus, the net electrical charge on the reactant side equals zero. On the product side, the electrical charges are:

$$2Fe_2^{+3} = 2 \times 2 \times (+3) = +12$$
$$2O_3^{-2} = 2 \times 3 \times (-2) = -12$$

The sum of -12 and $+12$ is zero; thus, the net electrical charge on the product side equals zero. In this reaction, Fe is oxidized and O is reduced. Half-reactions indicating the oxidation and reduction equations separately can be written as follows:

$4Fe^0\ (-12e^-) \rightarrow 2Fe_2^{+3}$ oxidation (loss of e^-)
$3O_2^0 \rightarrow 2O_3^{-2}\ (+12e^-)$ reduction (gain of e^-)

Thus, the number of e^- lost ($-12e^-$) equals the number of e^- gained ($+12e^-$). Each Fe atom changes its oxidation number from zero to $+3$, and each O atom changes its oxidation number from zero to -2. Note that elements increase their oxidation number (become more positive) when they are oxidized and decrease their oxidation number (become more negative) when they are reduced.

3

PROPERTIES OF WATER

Pure water is a colorless liquid that boils and becomes a gas at 100°C and freezes to solid ice at 0°C. It is the most abundant inorganic compound found on earth as well as the most abundant single component of living tissues. The first life forms undoubtedly evolved in water, and water remains to this day the natural habitat for many organisms. The internal environment of the body is a watery fluid in which ions and molecules are either dissolved or dispersed. Not only do the biochemical reactions of cells require the presence of water, but water also actively participates in many of these reactions and is formed as an end product in others. The unique chemical and physical properties of water account for the fact that all living systems that have successfully evolved on earth are essentially water-based systems.

WATER MOLECULE

The water molecule consists of 2 atoms of hydrogen joined to an atom of oxygen by covalent bonds. The molecule is V-shaped, as shown in Fig. 3-1. This shape and the *electronegativity* of the oxygen create a strongly *dipolar* molecule (see Chapter 2). The polarity of the water molecule accounts for the formation of molecular aggregates of water through hydrogen bonding and for the unusual solvent properties of water.

HYDROGEN BONDS AND THE PHYSICAL PROPERTIES OF WATER

The nature of hydrogen bonds is discussed in Chapter 2. In liquid water, each water molecule is probably hydrogen bonded to at least 3 other water molecules (Fig. 3-2). The molecular aggre-

Fig. 3-1. Two representation of the water molecule. The shared electrons are attracted more strongly to the electronegative oxygen, creating a dipole. The oxygen end is negatively charged; the hydrogen ends are positively charged. The negativity of the oxygen pole is further enhanced by the two unshared pairs of electrons (lone pairs).

gates pictured in Fig. 3-2 are actually in continuous motion in the liquid state, and the bonds are constantly being broken and re-formed. Nevertheless, the clusters of hydrogen-bonded water molecules produce a degree of internal cohesion in water that is responsible for the following physical properties:

specific heat

1. A high *specific heat*. The amount of heat necessary to raise the temperature of water is greater than would be expected, due to the extra energy required for breaking the hydrogen bonds. As a result of this capacity to absorb heat without appreciable fluctuations in its own temperature, water moderates the temperatures of the earth as well as the temperatures of living organisms.

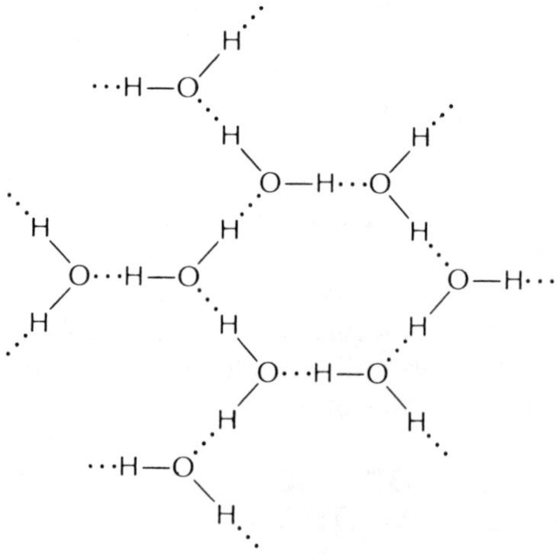

Fig. 3-2. Aggregates of water molecules hydrogen bonded to one another. This is a two-dimensional representation of a three-dimensional lattice. The hydrogen bonds are shown as dotted lines.

2. A high *heat of vaporization*. The amount of heat necessary to change water from a liquid to a gas is greater than would be expected. Thus, water that is evaporating carries away large quantities of heat. This is the main mechanism by which body temperatures in warm-blooded animals are stabilized.

heat of vaporization

3. A high *boiling point*. Without hydrogen bonds, water could be expected to boil at around −80°C. This temperature is so much below average environmental (15° to 20°C) and body (37°C) temperatures that living cells containing water could not possibly have evolved under such circumstances.
4. *Solid water (ice) is less dense than liquid water*. Each water molecule in ice is hydrogen bonded to 4 other water molecules to form an open lattice structure in which the molecules are rigidly held farther apart then the molecules in liquid water. As a result of these intermolecular spaces, ice floats on water and bodies of water always freeze from the top down, thus making it possible for a multitude of aquatic life forms to survive in spite of freezing air temperatures.

WATER AS A SOLVENT

The dipolar nature of the water molecule makes it an ideal solvent (dissolving medium) for a variety of substances.

1. Most ionic compounds dissolve readily in water. Ionic compounds are ionized even in the solid state; the oppositely charged ions are held together in the crystal by the electrostatic attractions between them (see Chapter 2). This interionic attraction is overcome when a solid ionic compound, for example, sodium chloride (Na^+Cl^-), is placed in water. The Na^+ ions are attracted to the negative ends of the water dipoles, and the Cl^- ions are attracted to the positive ends with the result that each Na^+ and Cl^- ion is surrounded by a shell of water molecules. The formation of stable *hydrated ions* in solution in water is shown in Fig. 3-3. In this case, the dissolving action of water permits the cations and anions of a solid ionic compound to move apart. This process is called *dissociation*. Ionic species in solution in body fluids exist as dissociated hydrated ions.

hydrated ions

dissociation

2. Covalent compounds with weakly polar properties are readily dissolved in water. The solubility of such organic compounds as sugars, simple alcohols, aldehydes, and ketones is due to the tendency of water molecules to form hydrogen

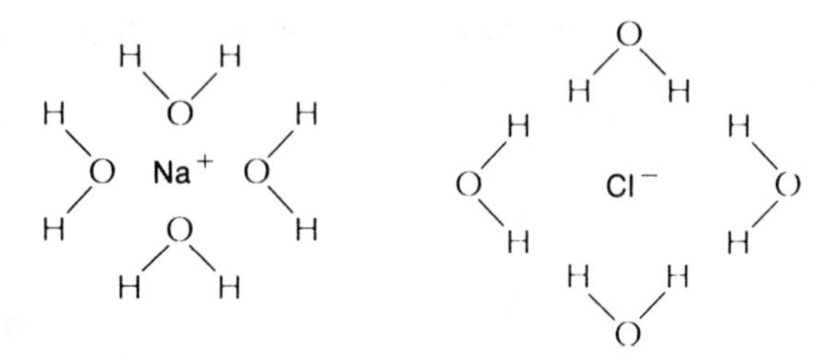

Fig. 3-3. Hydrated ions surrounded by shells of polar water molecules. The charged ends of the water molecules are attracted to the oppositely charged ions.

bonds with certain polar groups, such as the hydroxyl (OH) groups in sugars and alcohols and the carbonyl (C=O) groups in aldehydes and ketones.

It should be noted here that *nonpolar* covalent molecules, such as benzene, ether, chloroform and fats or oils, have little or no attraction for water dipoles and, as a result, are largely insoluble in water. Substances of this type are said to be *hydrophobic* (literally, to 'fear water'). We are all familiar with the saying that "oil and water do not mix." When hydrophobic molecules are placed in water, they associate closely with each other, and avoid associating with water. The forces that drive these characteristic arrangements are called *hydrophobic interactions*. Numerous hydrophobic compounds, many of them macromolecules, exist in the aqueous internal environment of the body, and hydrophobic interactions play an important role in determining their structure and activity. The solubilization of various types of lipid (fatty) substances in the aqueous body fluids is discussed in more detail in Chapter 6.

hydrophobic interactions

IONIZATION REACTIONS

Some non-ionized (covalent) compounds undergo a chemical reaction with water and form new species when they are dissolved. In these reactions ions are usually produced. When covalent molecules react with water to form ions, the process is called *ionization*. Technically, ionization differs from dissociation in that it is an actual chemical reaction in which nonionic substances are chemically changed to ions. However, the two terms are often used interchangeably.

ionization

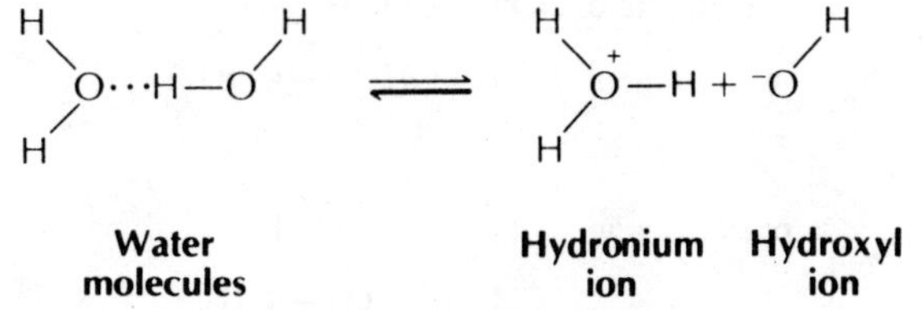

$$\begin{matrix} H & & H \\ & O\cdots H-O & \\ H & & \end{matrix} \rightleftharpoons \begin{matrix} H & & & H \\ & \overset{+}{O}-H + {}^{-}O & & \\ H & & & \end{matrix}$$

Water molecules — Hydronium ion — Hydroxyl ion

Fig. 3-4. Ionization of water.

Pure water may be considered a covalent compound that ionizes to a very slight extent. The interaction between water dipoles infrequently results in the complete transfer of a proton (a hydrogen nucleus, or *hydrogen ion,* H^+) from one water molecule to another (Fig. 3-4). The ions formed by the ionization of 2 water molecules are the *hydronium ion* (H_3O^+) and the *hydroxyl ion* (OH^-). The ionization of water is often shown more briefly as

hydrogen ion

hydroxyl ion

$$H_2O \rightleftharpoons H^+ + OH^-$$

although the use of the term *hydrogen ion* and its symbol H^+ is not strictly accurate. The hydrogen ion is a bare proton that does not exist independently in water; it is always hydrated, that is, bonded to a variable number of water molecules. This fact should be kept in mind when using the simpler and more convenient notation H^+ rather than the hydrated form H_3O^+.

The quantity of H^+ and OH^- formed in pure water is minute; approximately 1 hydrogen ion and 1 hydroxyl ion is formed for every 554 million nonionized water molecules. A liter of pure water at 25°C will contain 0.0000001 mole of H^+ ions and an equal amount of OH^- ions. This is expressed by the following chemical notations:

$$[H_3O^+] = 1 \times 10^{-7}\ \text{M (molar, or mole per liter)}$$

$$[OH^-] = 1 \times 10^{-7}\ \text{M}$$

(The use of square brackets in chemical shorthand indicates *concentration of.*)

Covalent compounds that react with water when dissolved and ionize to produce a concentration of hydronium ions *greater than* 10^{-7} M are known as *aqueous acids*. These substances include the inorganic acids—hydrochloric acid (HCl), sulfuric acid (H_2SO_4), nitric acid (HNO_3), phosphoric acid (H_3PO_4), carbonic acid (H_2CO_3)—and some organic acids, for example:

aqueous acids

acetic acid $CH_3—COOH$

lactic acid $CH_3—CHOH—COOH$

pyruvic acid $CH_3—\overset{\overset{\displaystyle O}{\|}}{C}—COOH$

citric acid $\begin{array}{ccc} CH_2— & CHOH— & CH_2 \\ | & | & | \\ COOH & COOH & COOH \end{array}$

proton donor

The common feature of all aqueous acids is that in solution they will donate 1 or more protons, that is, hydrogen ions, to molecular water to form hydronium ions.

carboxyl group

With reference to the organic acids listed above, it should be noted that the presence of hydrogen atoms in the molecule is not in itself sufficient to classify a compound as an acid. The compound must be able to *transfer* 1 or more protons. In most organic acid molecules, the only proton-donating group is the *carboxyl group* (COOH). The ionization reaction of acid molecules in water is shown as follows, using hydrochloric acid (HCl) and acetic acid (CH_3COOH) as examples:

1. $HCl + H_2O \rightleftharpoons \underset{\text{hydronium ion}}{H_3O^+} + \underset{\text{chloride ion}}{Cl^-}$

(abbreviated version) $HCl \rightleftharpoons \underset{\text{hydrogen ion (aqueous)}}{H^+(aq)} + \underset{\text{chloride ion (aqueous)}}{Cl^-(aq)}$

2. $CH_3COOH + H_2O \rightleftharpoons \underset{\text{hydronium ion}}{H_3O^+} + \underset{\text{acetate ion}}{CH_3COO^-}$

(abbreviated version) $CH_3COOH \rightleftharpoons \underset{\text{hydrogen ion (aqueous)}}{H^+(aq)} + \underset{\text{acetate ion (aqueous)}}{CH_3COO^-(aq)}$

acid strength

The distinction between *strong acids* and *weak acids* is made on the basis of the extent to which the acid ionizes, that is, its readiness to donate protons to water. Hydrochloric acid is a strong acid; it reacts completely with water to form the hydronium ion and the Cl^- anion. Acetic acid is a weak acid; that is, in solution the concentration of hydronium and acetate ions will be considerably less than nonionized (undissociated) molecular acetic acid. These relationships are indicated by the light and heavy arrows in the above chemical equations to suggest the relative amounts of reactants and products that would be present when chemical equilibrium is established.

Polyprotic acids can donate more than 1 proton per molecule. These acids include sulfuric acid (H_2SO_4), carbonic acid (H_2CO_3), and phosphoric acid (H_3PO_4). Polyprotic acids ionize in stages, losing 1 proton at a time, as shown:

1. $H_2SO_4 + H_2O \rightleftharpoons H_3O^+ + HSO_4^-$ (bisulfate ion)

   $HSO_4^- + H_2O \rightleftharpoons H_3O^+ + SO_4^{2-}$ (sulfate ion)

2. $H_2CO_3 + H_2O \rightleftharpoons H_3O^+ + HCO_3^-$ (bicarbonate ion)

   $HCO_3^- + H_2O \rightleftharpoons H_3O^+ + CO_3^{2-}$ (carbonate ion)

3. $H_3PO_4 + H_2O \rightleftharpoons H_3O^+ + H_2PO_4^-$ (dihydrogen phosphate ion)

   $H_2PO_4^- + H_2O \rightleftharpoons H_3O^+ + HPO_4^{2-}$ (monohydrogen phosphate ion)

   $HPO_4^{2-} + H_2O \rightleftharpoons H_3O^+ + PO_4^{3-}$ (phosphate ion)

When an acid donates a proton, it must be gained by a *base*. A base is a *proton acceptor*. When HCl and CH_3COOH are dissolved in water, water molecules act as bases, that is, as proton acceptors. Bases thus may be neutral molecules, such as water or ammonia (NH_3), or ions, such as acetate ion (CH_3COO^-). Like acids, bases differ in strength, that is, in their ability to accept protons. The hydroxyl ion, OH^-, is the strongest base that can exist in water.

base
proton acceptor

Substances that either contain the OH^- ion or ionize in water to form OH^- ions are generally classified as *aqueous bases*. These include ionic compounds such as sodium hydroxide (NaOH) and

aqueous bases

Table 3-1. Properties of aqueous acids and bases

Aqueous acids	Aqueous bases
Proton donors	Proton acceptors
pH less than 7 (water solutions acidic)	pH more than 7 (water solutions basic)
Taste sour	Taste bitter
	Feel soapy on the fingers
Turn blue litmus red	Turn red litmus blue

potassium hydroxide (KOH), which dissociate in water, as well as molecules such as ammonia (NH_3), which reacts with water to form the ammonium ion (NH_4^+), and the hydroxyl ion (OH^-). The strong alkali metal hydroxides (NaOH and KOH) are commonly called caustics or alkalis.* Aqueous bases produce a hydroxyl ion concentration in water that is greater than 10^{-7} M (mole per liter).

Aqueous acids and bases have certain distinctive properties (summarized in Table 3-1), among which is their ability to change the color of certain organic dyes known as indicators. The indicator litmus is a good example of a vegetable dye that shows different colors with acids and bases.

neutralization

When an aqueous base (containing OH^- ions) is added to an aqueous acid, a *neutralization* reaction occurs, in other words, the acid and base neutralize each other. Neutralizations are double replacement reactions in which the anions and cations of the acid and the base exchange places. The hydronium ions of the acid combine with the hydroxyl ions of the base to produce water (a neutral compound). At the same time, the remaining anions of the acid combine with the remaining (usually metallic) cations of the base, forming an ionic compound called a *salt*. A general equation for this type of reaction is:

salt

$$\text{Acid} + \text{Base} \rightarrow H_2O + \text{Salt}$$

More specifically, the formation of the salt, sodium chloride (NaCl), by a neutralization reaction between aqueous hydrochloric acid (HCl) and aqueous sodium hydroxide (NaOH) may be shown as follows:

$$HCl(aq) + NaOH(aq) \rightarrow H_2O + NaCl$$

It should be borne in mind that the acid, base and salt in the above equation actually exist in the form of ions, that is, H_3O^+, Na^+,

* The terms *basic* and *alkaline* are often used interchangeably.

OH^-, and Cl^-. Sodium chloride, which is common table salt, is in chemical terms the sodium salt of hydrochloric acid. It is only one of many hundreds of different salts known to chemists.

For the rest of this discussion, the term *hydrogen ion* (H^+) will be used instead of *hydronium ion*. The hydrogen ion is simpler to work with as long as one remembers that H^+ always exists in an aqueous medium as H_3O^+.

MEASUREMENT OF HYDROGEN IONS IN AQUEOUS SOLUTIONS: pH

The function of biological macromolecules, and particularly of the enzymes of the body, depends on the concentration of hydrogen ions $[H^+]$, in the body fluids. As has been indicated, $[H^+]$ and $[OH^-]$ are each equal to 10^{-7} M in pure water at 25°C. If the concentrations of the two ions are multiplied, a constant is obtained, which is called the *ion product* of water, or K_w. This constant has the value of 1.0×10^{-14}:

K_w

$$K_w = [H^+] \times [OH^-] = (1 \times 10^{-7}) \times (1 \times 10^{-7}) = 1 \times 10^{-14}$$

An acid or base in aqueous solution may change the concentration of either the hydrogen ions or the hydroxyl ions, but the constant ion product K_w cannot exceed 1×10^{-14}. This means that if the concentration of one ion increases, the concentration of the other must decrease, and vice versa. Since the concentrations of the two ions are so closely correlated, the hydrogen ion concentration alone is used to describe the characteristics of aqueous solutions; that is, whether they have more than 10^{-7} mole of hydrogen ions per liter and are acidic, or less than 10^{-7} (more than 10^{-7} hydroxyl ions) and are basic. The pH scale makes it possible to express hydrogen ion concentrations more concisely:

$$pH = -\log^* [H^+] = \log \frac{1}{[H^+]}$$

where,

$[H^+]$ = molar concentration (M) of hydrogen ions in a solution (mole per liter)

or,

$$[H^+] = 10^{-pH}$$

* The definition of the common logarithm, that is $\log_{10}$, as used in this expression, is the power to which 10 must be raised to equal the number in question.

Pure water is a precisely neutral solution where the hydrogen ion concentration is equal to 1×10^{-7} M. The pH of pure water is calculated as follows:

$$[H^+] = 10^{-7} \text{ M}$$

$$pH = \log \frac{1}{10^{-7}} = \log 10^7 = 7$$

or,

$$10^{-pH} = 10^{-7}$$

so,

$$-pH = -7, \; pH = 7$$

Neutrality, therefore, is represented by a pH of 7.

The pH range of aqueous solutions can vary between 0 and 14 (Table 3-2). As the negative exponents of 10 *decrease,* the hydrogen ion concentration *increases* (10^{-1} is a larger number than 10^{-2} or 10^{-3}; see appendix to Chapter 1). Thus, pH values of less than 7 indicate an acidic solution. As the negative exponents of 10 *increase,* the concentration of hydrogen ions *decreases* (and the concentration of hydroxyl ions increases). Thus, pH values

Table 3-2. pH scale of aqueous solutions

		pH	$[H^+]$ (mole per liter)	$[OH^-]$ (mole per liter)
Acidic	Strong	0	1.0 (10^0)	10^{-14}
	↑	1	0.1 (10^{-1})	10^{-13}
		2	0.01 (10^{-2})	10^{-12}
		3	0.001 (10^{-3})	10^{-11}
		4	10^{-4}	10^{-10}
		5	10^{-5}	10^{-9}
	Weak	6	10^{-6}	10^{-8}
	Neutral	7	10^{-7}	10^{-7}
Basic	Weak	8	10^{-8}	10^{-6}
		9	10^{-9}	10^{-5}
		10	10^{-10}	10^{-4}
		11	10^{-11}	0.001 (10^{-3})
		12	10^{-12}	0.01 (10^{-2})
	↓	13	10^{-13}	0.1 (10^{-1})
	Strong	14	10^{-14}	1.0 (10^0)

Table 3-3. Comparative pH values of various fluids

Pure water	7.0
Seawater	7.0–7.5
Body fluids	
Blood plasma	7.36–7.44
Interstitial (tissue) fluid (ISF)	7.4 (average)
Intracellular fluid (ICF)	6.9–7.3
Cerebrospinal fluid (CSF)	7.35–7.45
Body secretions	
Bile (gallbladder)	7.0–7.6
Gastric juice	1.2–3.0
Pancreatic juice	7.5–8.0
Saliva	6.4–7.0
Urine	4.5–8.0
Foods	
Vinegar	3.0
Lemon juice	2.3
Tomato juice	4.3
Cola drinks	2.8
Cow's milk	6.6

of more than 7 indicate basic or alkaline solutions. At the extreme acidic end of the pH range, where pH = 0, the concentration of hydrogen ions $[H^+]$ is 1.0 M. At the extreme basic end, where pH = 14, the concentration of hydroxyl ions $[OH^-]$ is 1.0 M. All intermediate values indicate fractions of 1.0 (1/10, 1/100, 1/1000, and so on). Note that the pH scale is logarithmic: *when two solutions differ by 1 pH unit,* it means that *one solution has ten times the hydrogen ion concentration of the other*. A list of pH values of some common fluids is given in Table 3-3.

ACID-BASE BALANCE AND BUFFERS

It can be seen from Table 3-3 that the pH of body fluids is close to neutrality (approximately pH 7). A moderately narrow range of blood pH, between the extremes of 6.8 to 7.8, is compatible with life, although in a purely chemical context, the weakly acidic pH 6.8 and the weakly basic pH 7.8 represent insignificant deviations from neutrality. Where the function of the body is concerned, blood pH values that are 0.1 to 0.2 points *below* 7.35 or *above* 7.45 are associated respectively with the conditions of *acidosis* and *alkalosis,* both of which are serious, potentially fatal disturbances. The two extremes of the pH range cited above could actually be tolerated only for short periods of time. It can be

acidosis
alkalosis

optimum pH

inferred from this that the maintenance of a relatively constant *optimum pH* in the blood and other body fluids is vitally necessary to survival. In fact, the biochemistry of life depends on it to the extent that protein molecules, *particularly enzymes* (see Chapter 9), as well as other pH-sensitive molecules, cease to function when the concentration of H^+ ions goes beyond specifically defined limits.

homeostasis

All living systems exhibit the phenomenon of *homeostasis* (*homeo,* the same; *stasis,* standing); that is, they have built-in mechanisms that enable them to maintain their internal environment in a relatively constant state in spite of the many changes that take place both inside and outside the system during normal day-to-day existence. (Such mechanisms are based on complex *negative feedback mechanisms.*) The term *homeostasis* does not imply "standing still," but a state of dynamic equilibrium that may be compared to the swing of a pendulum moving in a prescribed arc around a central point. Within reasonable limits, when the pendulum swings too far to one side or another, feedback mechanisms come into play to compensate or correct for the deviation and restore the status quo. The maintenance of *acid-base balance* in the body is but one of countless examples of the operation of this principle in living organisms.

negative feedback

acid-base balance

How a Buffer System Works

buffers

The addition of a few drops of concentrated acid or base to a test tube of pure water will produce marked changes in pH. However, if similar small amounts of acid or base are added to a sample of blood plasma, only very minor variations in pH will be observed. This stabilization of pH is due to the presence of *buffers,* which are, in the body, the first line of defense against any challenge to acid-base homeostasis. A *buffer is a chemical system that resists changes in pH within reasonable limits.* The buffers we shall discuss are a group of weak acids in solution in the aqueous medium of body water that are only *partially* ionized (dissociated) to form the hydrogen ion (proton) plus an anion, or conjugate base. The anion is called this because it is a proton (H^+ ion) acceptor; it is also often called the *salt* of the weak acid. The standard terminology for any buffer system is HA (the undissociated, or non-ionized, buffer acid)—H^+ (the hydrogen ion, or proton) and A^- (the anion, conjugate base, or salt of the acid). The standard chemical equation that represents the dissociation equilibrium of any buffer solution is:

1. Buffer Equilibrium

$$HA \rightleftharpoons H^+ + A^-$$

In considering this model, it is important to keep the following facts in mind:

1. With respect to the equilibrium, the concentrations of the three components of the buffer system are generally in a fixed ratio to each other. When a standard amount of acid or base is added to a buffered solution (this operation is called *titration* in chemistry), the equilibrium will be displaced to the right or to the left, as the case may be, to compensate within reasonable limits for changes in the concentration of any one of these components and the ratio will be maintained.
2. In titrating a buffer, we are mainly interested in possible changes that might occur in the concentration of the H^+ ion or, for convenience, the pH of the system. Thus, any added substance that *donates* free protons (or H^+ ions) to the buffer constitutes an *acidic* challenge, and any added substance that *removes* free protons (H^+ ions) from the buffer constitutes a *basic* challenge. In the discussion following, the addition of acid or base to a buffer will be represented by the symbols, $\textcircled{H^+}$ and $\textcircled{OH^-}$, respectively.
3. Although the different versions of the A^- base of a buffer will be treated as independent ions in the material that follows, it should be noted that A^- bases are normally found in ionic association with positively charged cations in the body fluids. These cations are "spectators" in the sense that they are present but undergo no chemical changes during the buffering reactions. In extracellular fluids, such as blood plasma, the predominant cation is Na^+ (sodium ion), and the A^- anion of the buffer in these fluids is often represented in its combined form as NaA (the sodium *salt*); in intracellular fluid, the predominant cation is K^+ (potassium ion), and the anion of an intracellular buffer is often given as KA (the potassium salt).

If a strong acid is added to a buffer, the $\textcircled{H^+}$ ions of the strong acid combine with the available A^- base of the buffer as follows:

2. Buffering of Excess Acid

$$\textcircled{H^+} + A^- \rightarrow HA$$

In effect, this decreases the concentration of A^-, and displaces the equilibrium (equation 1) to the *left*. By combining with A^- to form HA, excess free H^+ is removed from the solution because HA is a *weak* acid; that is, it tends to hold on to the protons it picks up from the solution.

If a strong base, denoted by the hydroxyl ion (OH^-), is added to a buffer, the basic OH^- ions take protons away from the buffer acid to yield H_2O:

3. Buffering of Excess Base

$$OH^- + HA \rightarrow H_2O + A^-$$

In the formation of H_2O, the excess OH^- base is neutralized. At the same time, the H^+ ion concentration of the solution does not appreciably decrease, because the equilibrium (equation 1) is displaced to the *right*. This means that more HA now dissociates and donates H^+ ions to make up for the deficit and reestablishes the equilibrium.

These shifts in the dissociation equilibrium of a buffer system constitute the mechanism by which changes in pH are minimized. Furthermore, the dynamics of these reactions can be expressed mathematically, so that the concentrations of the H^+ ion (and the pH), the undissociated buffer, and its anion in any given set of conditions can be quantitatively estimated. Buffers generally operate with equal effectiveness *in vitro* (in the test tube) and *in vivo* (in the living body), and thus have important applications in chemistry as well as biology. In this text, however, we will deal only with the major physiological buffer systems of the body fluids.

Carbonic Acid-Bicarbonate Buffer System (H_2CO_3/HCO_3^-)

H_2CO_3/HCO_3^- system

The carbonic acid–bicarbonate buffer system, a major buffer system of blood, consists of carbonic acid (H_2CO_3) and its base, the bicarbonate anion (HCO_3^-). The equilibrium of these components is as follows:

$$H_2CO_3 \rightleftharpoons H^+ + HCO_3^-$$

The bicarbonate system is a specialized one in that the H_2CO_3 of blood is derived from the gas carbon dioxide (CO_2), which is produced continuously by cells as a waste product of cell metabolism and then blown off by the lungs. The chemical reaction by which dissolved CO_2 is hydrated in H_2O to form H_2CO_3 is normally very slow. It is speeded up many thousands of times by the enzyme *carbonic anhydrase,* which is present in red blood cells:

carbonic anhydrase

$$CO_2 + H_2O \xrightleftharpoons[\text{anhydrase}]{\text{carbonic}} H_2CO_3$$

The entire equilibrium for the bicarbonate buffer system, with all the components involved, can thus be shown in this manner:

Buffer Equilibrium

$$H_2O + CO_2 \rightleftharpoons H_2CO_3 \rightleftharpoons H^{+*} + HCO_3^-$$

It will be immediately evident from the equation that there must be a very close correlation between the concentration of H_2CO_3 and the amount of CO_2 dissolved in the blood (this is proportional to the partial pressure of CO_2 in the blood, expressed by the term P_{CO_2}). The unique feature here is that the carbonic acid buffer is in equilibrium with a gas that can be excreted by the lungs to a greater or lesser extent, depending on whether the subject breathes faster or slower. Since the *rate* of respiration is primarily controlled by unconscious centers in the brain stem that are very sensitive to fluctuations in the P_{CO_2} and the pH of blood, respiratory feedback plays a key role in the operation of this buffer system. In fact, the carbonic acid-bicarbonate buffer is very much more effective *in vivo* than *in vitro,* precisely because it is controlled by respiratory mechanisms. At the prevailing pH 7.4 of arterial blood, the normal ratio of base to acid (or dissolved CO_2†) is approximately 20:1, that is:

$$\frac{[HCO_3^-]}{[H_2CO_3]} = \frac{24 \text{ mmole/liter}}{1.2 \text{ mmole/liter}} = \frac{20}{1} = \text{pH } 7.4$$

Note that in venous blood, which has a somewhat higher CO_2 content than arterial blood, this ratio is about 19:1 at a pH of 7.38. Buffers generally exhibit maximum buffering capacity at pH levels at which the *ratio* of base to acid (and not the absolute quantities of each substance) is approximately 1:1. The reason the bicarbonate system functions effectively in the blood with a

* The H^+ ions generated by this reaction in red blood cells are buffered by hemoglobin (see pp. 62–63).

† P_{CO_2} is measured in *torr* (mm Hg), which can be converted to mmole/liter (1 mmole = 10^{-3} mole). Thus, the P_{CO_2} of arterial blood is 40 torr, equivalent to 1.2 mmole/liter, and the P_{CO_2} of venous blood is 46 torr, equivalent to 1.37 mmole/liter.

ratio of 20:1 at the prevailing pH 7 range is because blood H_2CO_3 can be rapidly replenished by supplies of metabolically-generated CO_2.

If an acid is added to the blood, the excess H^+ ions are removed by the bicarbonate base (HCO_3^-) to form carbonic acid (H_2CO_3); and as the concentration of H_2CO_3 builds up, it decomposes to H_2O and CO_2. The buffer equilibrium above is thus displaced to the *left:*

$$H^+ + HCO_3^- \rightarrow H_2CO_3 \rightarrow H_2O + CO_2 \uparrow$$

As the excess CO_2 is blown off by the lungs, the buffer equilibrium is reestablished. The rule here is: When an acid challenge threatens to lower the pH of blood (raise the H^+ ion concentration), it can be compensated for by an *increase* in the respiratory rate.

If excess base, indicated by the OH^- ion, appears in the blood, the base will remove protons from H_2CO_3 to form H_2O and the bicarbonate ion:

$$OH^- + H_2CO_3 \rightarrow H_2O + HCO_3^-$$

In the process of being neutralized to H_2O, the base uses up carbonic acid. However, any carbonic acid deficit can readily be made up by allowing P_{CO_2} to build up again in blood:

$$CO_2 + H_2O \rightarrow H_2CO_3$$

Once again, the buffer equilibrium is effectively maintained by respiratory mechanisms. The rule here is: When a basic or alkaline challenge threatens to raise the pH of blood (lower the H^+ ion concentration), it can be compensated for by a *decrease* in the respiratory rate.

$H_2PO_4^-/HPO_4^{2-}$ system

Phosphate Buffer System ($H_2PO_4^-/HPO_4^{2-}$)

The weak acid, or proton donor, of the phosphate buffer system is the more acidic of the two phosphate ions—the dihydrogen phosphate ion ($H_2PO_4^-$)—and the base, or proton acceptor, of the system is the monohydrogen phosphate ion (HPO_4^{2-}). Note that in body fluids, phosphoric acid (H_3PO_4) ionizes to form these two ions (see p. 51). The phosphate buffer is predominantly an intracellular buffer, although there is some phosphate buffer in blood. At a pH slightly below 7, the ratio of base to acid, $HPO_4^{2-}/H_2PO_4^-$, is about 1:1; this buffer system therefore has maximum buffering capacity mainly at intracellular pH levels. In the blood at pH 7.4, this ratio is approximately 5:1 and the buffer is correspondingly less effective. The equation for the equilibrium of this system is as follows:

Buffer Equilibrium

$$H_2PO_4^- \rightleftharpoons H^+ + HPO_4^{2-}$$

The base of the phosphate buffer, HPO_4^{2-}, readily buffers acid that is added to the system by combining with the added H^+ ions to form the weak acid ion, $H_2PO_4^-$:

$$H^+ + HPO_4^{2-} \rightarrow H_2PO_4^-$$

This displaces the equilibrium to the left.

When base is added to the buffer, as indicated here by OH^- ions, protons will be removed from the $H_2PO_4^-$ component of the buffer, to yield H_2O:

$$OH^- + H_2PO_4^- \rightarrow H_2O + HPO_4^{2-}$$

This shifts the equilibrium to the right, as more $H_2PO_4^-$ dissociates to make up the deficit of H^+ ions and stabilize the equilibrium.

The presence of phosphate buffer in the cells that line the renal tubules provides one of the important mechanisms by which the kidney compensates for an excess or deficit of H^+ ions in the body. The two phosphate ions are excreted in urine. When the kidney is conserving H^+ for the body, the base HPO_4^{2-} is the predominant phosphate ion found in the urine; when the kidney has to unload an excess of H^+, a more acidic urine containing $H_2PO_4^-$ will be excreted.

Protein Buffers (HProt/Prot$^-$); Hemoglobin as a Buffer (HHb/Hb$^-$)

Proteins (including hemoglobin, a specialized iron-containing protein found only in red blood cells) are the major organic* buffers of blood and intracellular fluid. Proteins are large, complex heterogeneous molecules, made up of varying numbers of amino

* Although carbonic acid and the bicarbonate ion are, by strict definition, organic compounds (containing carbon), they are conventionally grouped with other inorganic acids and bases.

acids, with certain characteristic ionizable acidic and basic groups on their side chains (see Chapter 7). They make effective buffers because they can act either as weak acids (proton donors) or as weak bases (proton acceptors) over a wide range of pH values. The characteristic buffering reactions of hemoglobin and the plasma proteins of blood exemplify the properties of protein buffers in general. At the pH 7.4 of blood, the plasma proteins and hemoglobin act as weak acid-base pairs, usually symbolized by the abbreviations *HProt* and *Prot*$^-$, and *HHb* and *Hb*$^-$, respectively. We shall discuss the plasma proteins first.

HProt/Prot$^-$ and HHb/Hb$^-$ buffers

The equilibrium for the dissociation of the weak acid buffer HProt into an H^+ ion and a base $Prot^-$ is indicated as follows:

Buffer Equilibrium

$$HProt \rightleftharpoons H^+ + Prot^-$$

Acid H^+ ions added to this system will be picked up by the anionic groups of the protein molecule ($Prot^-$), thus displacing the above equilibrium to the left:

$$H^+ + Prot^- \rightarrow HProt$$

In the process, the excess H^+ is removed from the solution, since HProt is only very weakly dissociated.

The addition of base (indicated by OH^-) to this system will result in the removal of the H^+ ions from the protein and in the formation of (neutral) H_2O:

$$OH^- + HProt \rightarrow H_2O + Prot^-$$

The equilibrium will now shift to the right, as more acidic groups on the protein molecule ionize, contributing additional H^+ ions to make up for the deficit and restoring the equilibrium.

The buffering activity of red blood cells is highly efficient because they contain hemoglobin. Because of the nature of certain of the side chains in the hemoglobin molecule, hemoglobin is a more effective buffer than the plasma proteins at the pH range of blood. Using the conventional abbreviations for the weak acid and the base, HHb/Hb^-, the dissociation equilibrium for this system is as follows:

Buffer Equilibrium

$$HHb \rightleftharpoons H^+ + Hb^-$$

Some of the side chains in the hemoglobin molecule have a great affinity for H^+, and their presence is responsible for the main function of hemoglobin as a buffer: *to neutralize the H^+ ions that are produced* when the following reaction takes place in red blood cells (see p. 59).

$$H_2O + CO_2 \xrightleftharpoons[\text{anhydrase}]{\text{carbonic}} H_2CO_3 \rightleftharpoons H^+ + HCO_3^-$$

The H^+ ions are removed by the Hb^- base, which is thereby converted to the weakly dissociated HHb form:

$$H^+ + HCO_3^- + Hb^- \rightarrow HHb + HCO_3^-$$

Note that the HCO_3^- ions (bicarbonate) are turned loose by this reaction. They then pass out of the red blood cells into the plasma. To maintain electrical neutrality, as HCO_3^- ions leave the red blood cells, Cl^- ions move into the cells from the plasma. This phenomenon is known as the *chloride shift* in respiratory physiology.*

chloride shift

A neat feature of the hemoglobin buffer system is the fact that unoxygenated hemoglobin (HHb) is a weaker acid than oxygenated hemoglobin (usually abbreviated as $HHbO_2$), and it is therefore a better buffer; that is, it has a greater capacity to mop up H^+ ions. Thus, as blood flows through the tissues of the body, hemoglobin simultaneously unloads its oxygen and increases its buffering power to cope with the incoming load of carbonic acid. The reverse of this process occurs in the lungs; at the same time as the stronger buffer, HHb, is oxygenated by the air in the lungs to the weaker buffer, $HHbO_2$, carbonic acid is unloaded and blown off in the form of CO_2.

Some Functional Guidelines

A full discussion in depth of normal and abnormal acid-base physiology is beyond the scope of this text. However, a few important points are briefly summarized here as an aid to students who wish to gain a more complete understanding of the fundamentals.

It should be noted that the buffer systems present in any particular body fluid act cooperatively, so that when an influx of acid or base challenges the pH balance of the fluid, the load is shared by all. This is shown diagrammatically below with respect to an

* When hemoglobin combines with O_2 in the lungs, the shift is reversed; that is, Cl^- ions move back into the plasma.

influx of excess H^+, for example (assuming that all the buffers are present and fully operative):

$$\text{excess } H^+ \begin{cases} HCO_3^- \rightarrow H_2CO_3\ (\rightarrow H_2O + CO_2 \uparrow) \\ HPO_4^{2-} \rightarrow H_2PO_4^- \\ Prot^- \longrightarrow HProt \\ Hb^- \longrightarrow HHb \end{cases}$$

Normally, the body has to cope with an excess of acid rather than an excess of base. There are several reasons for this. First, the bulk of the CO_2 given off as a waste product of metabolism by all cells is converted to H_2CO_3, which contributes H^+ ions to the body fluids. Second, other acids, such as sulfuric acid, phosphoric acid, and various organic acids, are produced during the course of normal metabolism in the body. Finally, muscular exertion is accompanied by the release of lactic acid (see Chapter 10). In sum, a large excess of acid is generated by normal day-to-day chemical functions in the body. Were it allowed to accumulate, the pH of body fluids could drop to about 0.5.

As noted previously, buffers are the first line of defense of acid-base balance in the body. The buffers are backed up in turn by a series of beautifully integrated feedback mechanisms involving the lungs and the kidneys. The slightly alkaline physiological pH is thus maintained by mechanisms at three levels: (1) buffer action, (2) respiratory responses controlling P_{CO_2} and $[H_2CO_3]$, and (3) renal responses controlling the excretion of acid and base in the urine. It is only when these mechanisms are partially or wholly overwhelmed that the clinical conditions of acidosis or alkalosis become evident.

ELECTROLYTES

Since ions are charged species, they can conduct an electric current. The conduction of electricity involves the movement of ions toward oppositely charged electrodes. *Electrodes* are usually pieces of metal that are connected by wires to a source of electricity such as a battery or a generator. The battery or generator causes electrons to move from one piece of metal to the other, so that one electrode has an excess of electrons (the negatively charged electrode, the *cathode*) and the other has a deficiency of electrons (the positively charged electrode, the *anode*).

electrodes

cathode

anode

When an electric current is passed through a solution of ions, the cations (positively charged ions) migrate to the cathode, where they combine with electrons to make up their electron deficit, and the anions (negatively charged ions) migrate to the anode, where they give up their excess electrons. (This give-and-take electron exchange is essentially the same process that occurs in oxidation-reduction reactions.)

Solid ionic substances cannot conduct electricity, because ions must be relatively free to move to the oppositely charged electrode. The best conductors are, therefore, species that have either dissociated, or ionized, when dissolved in water. Such substances are called *electrolytes*.

electrolytes

Most soluble ionic compounds are strong electrolytes. Molecular compounds, particularly acids that react completely with water to give a hydrogen ion and a corresponding anion, are strong electrolytes. Species that react slightly with water to form only a few ions are necessarily weak electrolytes. Species that form minute numbers of ions or no ions at all in solution, such as pure H_2O, sugars, and alcohols, do not conduct electricity and are called *nonelectrolytes*. It is of interest to note that electrolytes are not necessarily small inorganic chemical species; large organic biological molecules, such as proteins, that have acquired a net charge at certain pH levels, will also migrate in an electrical field toward the oppositely charged electrode. Dissolved electrolytes in the body water conduct electricity *in vivo* in much the same manner as they do in a laboratory container *(in vitro)*. The techniques of electrocardiography (ECG) and electroencephalography (EEG) are examples of *in vivo* conductivity. The source of electricity here is the small voltage generated by functioning muscle (the heart) and nerve tissue (the brain). Another, and rather grim, example is the fact that people can be shocked or electrocuted when their body fluid acts as a conductor for a large amount of electricity coming from an outside source.

The biological importance of electrolytes is based to a large extent on how these substances modify the properties of water. Body water is not pure H_2O; a great many substances, or *solutes*, are dissolved in it. When *one* molecule of a nonelectrolyte (nonionizable solute) is dissolved in water, it contributes *one* solute particle to the water. However, when one unit of an electrolyte is dissolved in water, it will dissociate, or ionize, to form 2 or more ions, *each of which* counts as a solute particle. Thus, electrolytes have a much greater effect on the properties of water than equivalent amounts of nonelectrolytes. The physical properties of water that are modified by dissolved electrolytes are termed *colligative properties*. These include the boiling point (elevated), the freezing point (depressed), and the osmotic pressure (increased) of water. These effects, particularly the osmotic effects (see Chapter 9), are profoundly significant in all biological sys-

solutes

colligative properties

Table 3-4. Principal electrolytes of blood plasma and intracellular fluid: approximate concentration in mmole per liter

	Plasma	Intracellular fluid
Cations		
Sodium (Na^{+})	145*	10–20
Potassium (K^{+})	4	140–160*
Calcium (Ca^{2+})	2	1
Magnesium (Mg^{2+})	0.7	15
Anions		
Chloride (Cl^{-})	110	4
Bicarbonate (HCO_3^{-})	26	10–15
Phosphate (HPO_4^{2-})	1	50
Sulfate (SO_4^{2-})	0.5	10
Protein ($Prot^{-}$)	2	8

* Note the characteristic pattern: sodium outside cells, potassium inside.

solvent

tems. Furthermore, electrolytes alter the dissolving *(solvent)* properties of water, thus influencing the solubility of large biological molecules. Considering the multitude of effects they have, merely as solutes, it might be said that electrolytes make up the physicochemical skeleton of body water.

Finally, electrolytes play many vital and necessary roles in the structural and functional biochemistry of the body. The importance of these substances to survival is emphasized by the variety of homeostatic mechanisms that have evolved in living organisms to maintain their concentration within strictly defined limits. A list of the major electrolytes of body fluids is given in Table 3-4; note that they include the buffer ions discussed previously.

4

CARBON COMPOUNDS

The chemistry of carbon compounds is called *organic chemistry*. Carbon atoms characteristically form covalent bonds with other carbon atoms, as well as with other elements. Bonds of this type make possible an enormous variety of different substances, both naturally occurring and synthetic. Carbon compounds are the common building blocks of all living organisms from the simplest unicellular forms to the most complex.

organic chemistry

CARBON BONDS

Carbon is an abundant nonmetallic element that is classified by its atomic structure in group IVA of the periodic table. The atomic number of carbon is 6, and the mass number of its most common isotope is 12 ($^{12}_{6}C$). The carbon atom has two main energy levels: the innermost shell, complete with 2 electrons, and the outermost valence shell, incompletely filled with 4 electrons (see Fig. 1-2). The electron dot structure for an atom of carbon with its 4 valence electrons is:

In forming compounds, carbon atoms show no tendency to gain or lose their 4 electrons; instead, they consistently form 4 covalent bonds with other atoms to attain a stable octet configuration. For convenience, the 4 covalent bonds that 1 carbon atom is capable of forming are represented as:

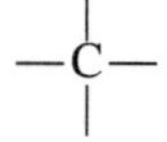

Each line represents a pair of shared electrons, one of the pair belonging to the carbon atom and the other electron belonging to another atom (not shown) to which the C is bonded. A diamond crystal, which is pure carbon, consists of many carbon atoms, each covalently bonded to 4 other carbon atoms to form a single large stable molecule:

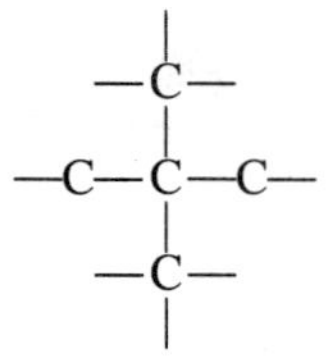

```
       |
      —C—
       |
  —C—C—C—
       |
      —C—
       |
```

double bond

Two carbon atoms can share two pairs of electrons with one another to form a *double bond*

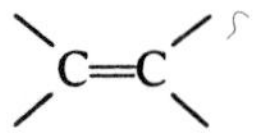

```
 \     /
  C═C
 /     \
```

triple bond

or three pairs of electrons to form a *triple bond.*

—C≡C—

Note that the above formulas represent incomplete molecules; in the double-bonded carbons *two* positions on each carbon atom are still open for bonding with other atoms, and in the triple-bonded carbons *one* position on each carbon atom is still open for bonding with other atoms. Hydrogen is very often found sharing its 1 valence electron with a carbon atom. Thus, in the above incomplete structural formulas the appropriate number of hydrogen atoms could be added to complete the molecules. A single carbon atom can share its 4 electrons with 4 hydrogen atoms to form the simple organic compound *methane:*

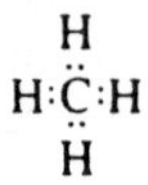

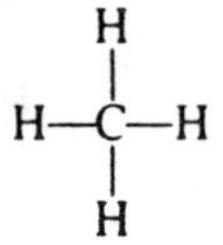

```
                    H
   H                |
H:C̈:H           H—C—H          CH4
   ̈                 |
   H                H
```

CH_4

The 2 double-bonded carbons can attain their stable octet configuration by forming bonds with 4 hydrogen atoms; this compound is *ethylene:*

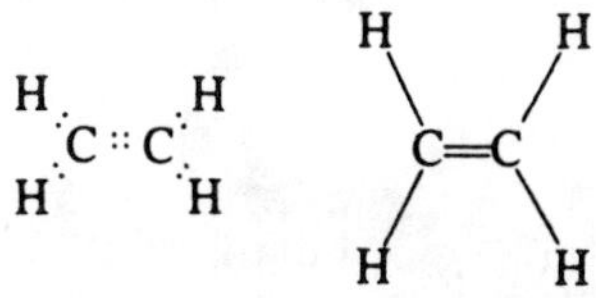

```
H     H           H       H
 .   .             \     /
 .C :: C.           C = C        CH2=CH2     C2H4
 .   .             /     \
H     H           H       H
```

$CH_2{=}CH_2$ C_2H_4

The 2 triple-bonded carbon atoms can each form a bond with hydrogen; this compound is *acetylene:*

H:C:::C:H H—C≡C—H CH≡CH C_2H_2
H—C≡C—H

The oxygen atom has 6 valence electrons and requires 2 more for a stable configuration. To attain this, it can share two pairs of electrons with carbon, forming a double bond, as in the compound *formaldehyde:*

```
  O
  ‖
  C            CH2O
 / \
H   H
```

CH_2O

The >C=O group is known as the *carbonyl group*.

carbonyl group

Oxygen can also form a single bond with a carbon atom and a second single bond with a hydrogen atom, as in the compound *methyl alcohol:*

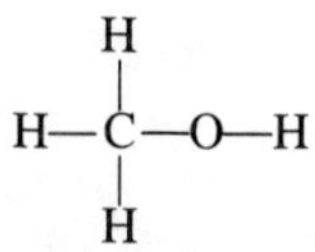

```
    H
    |
H—C—O—H     CH3OH
    |
    H
```

CH_3OH

hydroxyl group

The —OH group is called the *hydroxyl group*. Carbon and nitrogen can also form a single covalent bond:

```
  |
—C—N—
  |  |
```

The open-bond positions above can be filled by hydrogen atoms, as in the compound *methylamine:*

```
    H
    |
H—C—N—H     CH3NH2
    |  |
    H  H
```

structural formula — molecular formula (CH_3NH_2)

amino group

The $—NH_2$ group is called the *amino group*.

The carbonyl ($>C=O$), hydroxyl (—OH), and amino ($—NH_2$) groups are important features of many biological organic molecules.

Carbon can form single covalent bonds with the halogens fluorine, chlorine, bromine, and iodine (F, Cl, Br, I). These nonmetals all have 7 valence electrons. They acquire a stable octet configuration by sharing one pair of electrons in a covalent bond with the carbon atom. Two well-known compounds of this type are the anesthetic *chloroform*

```
      H
      |
Cl—C—Cl      CHCl3
      |
     Cl
```

structural formula — molecular formula ($CHCl_3$)

and the dry-cleaning solvent *carbon tetrachloride*.

```
     Cl
      |
Cl—C—Cl      CCl4
      |
     Cl
```

structural formula — molecular formula (CCl_4)

THREE-DIMENSIONAL CARBON ATOM

Carbon's four bonds are normally arranged equidistantly around the C atom. This is readily visualized by considering a ball-and-stick model, in which the carbon is a ball with four equidistant holes in its surface. When four equal sticks (representing the four bonds) are inserted into the holes, it becomes obvious that the bonds are equivalent; no matter which way the model is turned, it always stands on three legs and the fourth always points straight up (Fig. 4-1). This arrangement is the *tetrahedral* structure discussed briefly in Chapter 2. It is so called because the legs point to the corners of the tetrahedron, a three-dimensional solid made up of four triangular sides, with the carbon atom at its center.

tetrahedral carbon atom

Continuing with the ball-and-stick model, four balls can be attached to the four stick bonds, producing a general representation of a carbon atom bonded to four other groups. If these groups are all the same, or even if only two of them are alike (that is, three different groups) (Fig. 4-2, *A*), then the central carbon atom is symmetrical. This means that it is possible to pass a plane through the carbon and divide the molecule into symmetrical mirror-image halves. However, if the carbon is bonded to *four different groups,* it is *asymmetrical* (Fig. 4-2, *B*). There is no way to pass a plane of symmetry through such a carbon atom.

asymmetrical carbon atom

If the groups around a symmetrical carbon atom are rearranged, then nothing has changed; the two models are completely equivalent. The new model can simply be rotated, and it will be superimposable on the old (Fig. 4-3, *A*). However, if four different groups around an asymmetrical carbon are rearranged, then the new molecule will be three-dimensionally different from the original. For example, with group A pointing up, groups B, C, and D (clockwise) form the base. If groups B and C exchange places, there is no way that the new model can be rotated to make it

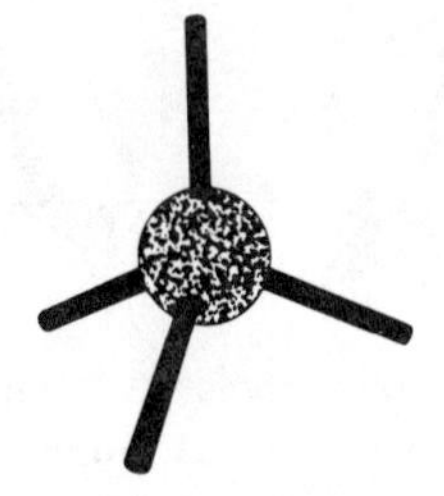

Fig. 4-1. Ball-and-stick model of a carbon atom and its four equivalent bonds.

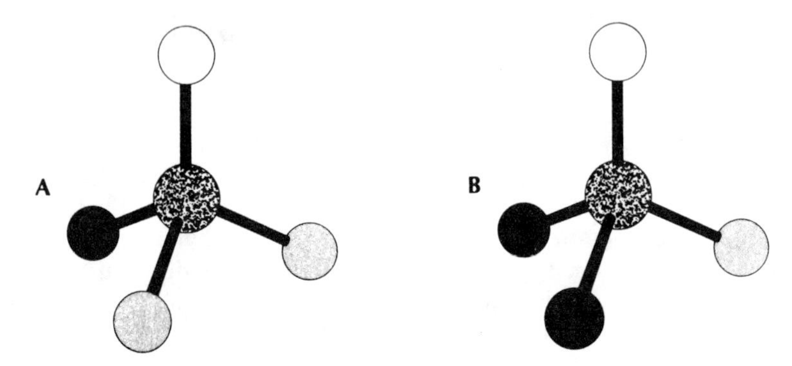

Fig. 4-2. Ball-and-stick models of carbon compounds. **A,** Symmetrical molecule with carbon atom (center) bonded to three different groups; a vertical plane of symmetry can be passed through the center of the carbon (between the two light-colored groups), dividing the molecule into mirror-image halves. **B,** Asymmetrical molecule; the carbon atom is bonded to four different groups.

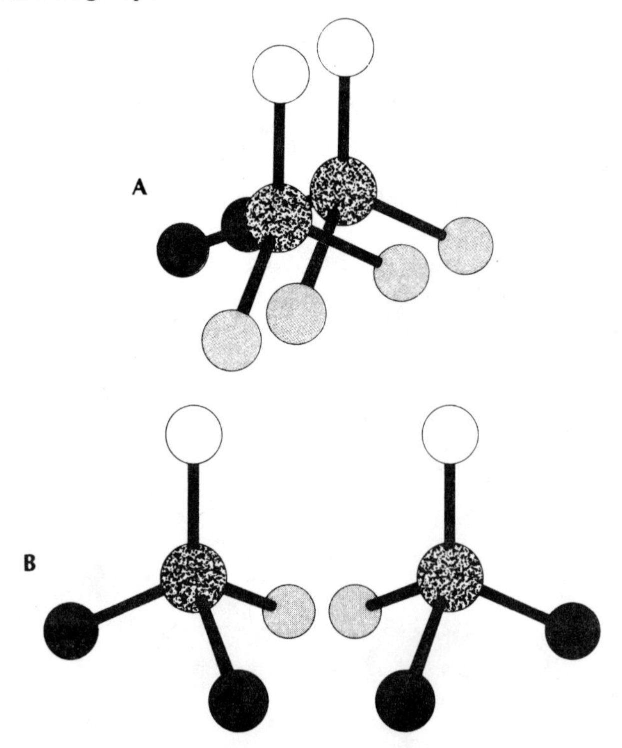

Fig. 4-3. Comparison of carbon atoms. **A,** Symmetrical—rearrangement of groups around 1 carbon has no effect. Superimposition is still possible simply by rotating the model. **B,** Asymmetrical—rearrangement of groups around 1 carbon atom produces a mirror image, which cannot be superimposed on the original.

superimposable on the original; in fact, they will be mirror images (Fig. 4-3, *B*), like a person's right and left hands.

In strictly chemical systems, this three-dimensional difference means little; however, in biological systems it is critical. Biological systems differ from mere chemical systems in that the thousands of chemical reactions are closely regulated as to which ones occur, how fast, when, and where. One important facet of this regulation is the recognition of three-dimensional structures by large biomolecules, especially enzymes (see Chapter 8). Generally, if a molecule has an asymmetrical carbon and can thus exist in two different forms, only one of those forms will be active in biological systems. This concept is discussed again in relation to sugars (see Chapter 5) and amino acids (see Chapter 7).

HYDROCARBON CHAIN COMPOUNDS

When carbon atoms form bonds with each other and with hydrogen atoms, organic compounds called *hydrocarbons* are formed. Hydrocarbon molecules may exist as *chain* or *ring* structures. Hydrocarbons that involve only single-bonded carbon atoms are called *saturated hydrocarbons;* those that involve double or triple bonds between carbon atoms are *unsaturated hydrocarbons*.

hydrocarbons

saturated
unsaturated

Saturated Hydrocarbon Chains: The Alkanes

Saturated hydrocarbon chains belong to a homologous series of compounds called the *alkanes*. The simplest alkane is the gas methane (CH_4). If the carbon chain is increased by successive —CH_2— units, a group of compounds with the general formula C_nH_{2n+2} results. The alkanes start with 1 carbon atom; they could have a thousand carbon atoms. The formation of the first few compounds of this series is shown in Table 4-1.

alkanes

Isomerism. Two structures are possible for the 4-carbon alkane butane. Both have the molecular formula C_4H_{10}.

```
                                  H
                                  |
    H   H   H   H              H—C—H
    |   |   |   |              H  |  H
 H—C—C—C—C—H                 |  |  |
    |   |   |   |           H—C—C—C—H
    H   H   H   H              |  |  |
                               H  H  H
```

normal butane   isobutane

Table 4-1. Saturated chain hydrocarbons: alkane series (C_nH_{2n+2})

Name	Empirical and molecular formulas	Structural formula
Methane (n = 1)	CH_4	H—C(H)(H)—H
Ethane (n = 2)	C_2H_6 CH_3CH_3	H—C(H)(H)—C(H)(H)—H
Propane (n = 3)	C_3H_8 $CH_3CH_2CH_3$	H—C(H)(H)—C(H)(H)—C(H)(H)—H
Butane (n = 4)	C_4H_{10} $CH_3CH_2CH_2CH_3$	H—C(H)(H)—C(H)(H)—C(H)(H)—C(H)(H)—H

structural isomers

Normal butane and isobutane are two different compounds. They are said to be *structural isomers* of each other; that is, they have the same molecular formula but different structural formulas. Structural isomerism is a common property of organic molecules and is one of the factors responsible for the great variety of organic compounds. The number of possible three-dimensional arrangements of organic molecules increases with their size and complexity.

alkyl groups

Alkyl (R) groups. An alkane can enter into combination with other atoms or groups of atoms when a hydrogen atom is removed from a molecule, leaving an open bonding position. Such groups formed from the alkanes are called *alkyl groups*. For example, CH_3— (the methyl group) would be the alkyl group for methane. The general chemical symbol for an alkyl is R—. Table 4-2 lists a few of the more common alkyl groups.

Table 4-2. Some alkyl groups

Name	Formula (abbreviated version)	
General group symbol	R—	
Methyl group	CH_3—	
Ethyl group	CH_3CH_2—	(C_2H_5—)
Propyl group	$CH_3CH_2CH_2$—	(C_3H_7—)
Isopropyl group	$CH_3CH(CH_3)$—	(C_3H_7—)
Butyl group	$CH_3CH_2CH_2CH_2$—	(C_4H_9—)
Isobutyl group	$CH_3CH(CH_3)CH_2$—	(C_4H_9—)

Unsaturated Hydrocarbon Chains: Alkenes and Alkynes

An added element of variation in carbon compounds is introduced by the presence of double and triple carbon-to-carbon bonds. Unsaturated hydrocarbons that have a double bond are called *alkenes;* those containing a triple bond are *alkynes.* These compounds form the same general pattern of homologous series as the alkanes.

alkenes
alkynes

The simplest compound in the alkene series is *ethylene* (C_2H_4):

$$H_2C{=}CH_2$$

The general formula for this series is C_nH_{2n}.

The simplest compound in the alkyne series is *acetylene* (C_2H_2):

$$H{-}C{\equiv}C{-}H$$

The general formula for the alkynes is C_nH_{2n-2}.

HYDROCARBON DERIVATIVES

The addition of oxygen or nitrogen groups to hydrocarbon molecules produces a variety of commonly occurring hydrocarbon derivatives that may be classed as *alcohols, aldehydes and ke-*

Table 4-3. Common derivatives of hydrocarbons

Added group	Characteristic structure	Class of compound formed	Class formula	Examples
Hydroxyl (OH) group	—C—OH	**Alcohol**	R—OH	Methyl alcohol (methanol): CH_3OH Ethyl alcohol (ethanol): C_2H_5OH
	(—C—OH)$_n$	**Polyhydroxy alcohol***		Glycerol: $CH_2(OH)—CH(OH)—CH_2(OH)$
Carbonyl (C=O) group at *end* of carbon chain	—C(H)=O	**Aldehyde**	R—CHO	Formaldehyde: HCHO Acetaldehyde: CH_3CHO
Carbonyl (C=O) group at *intermediate* position in chain	—C—C(=O)—C—	**Ketone**	R—C(=O)—R	Acetone: CH_3COCH_3
Carboxyl (COOH) group (carbonyl group plus hydroxyl group)	—C(=O)—OH	**Organic (carboxylic) acid†**	R—C(=O)—OH	Acetic acid: CH_3COOH (vinegar)
Carboxyl group plus amino (NH_2) group	—C(H)(NH_2)—C(=O)—OH	**Amino acid‡**	R—C(H)(NH_2)—C(=O)—OH	Glycine (simplest amino acid): $NH_2—CH_2—COOH$
Chemical combination of an acid and an alcohol:	—C(=O)—O(H + HO)—C— → H_2O + —C(=O)—O—C—	**Ester§**	R—C(=O)—O—R	Fats and oils are esters of glycerol and fatty acids
Chemical combination of two alcohols:	—C—O(H + HO)—C— → H_2O + —C—O—C—	**Ether**	R—O—R	Diethyl ether (anesthetic): $C_2H_5—O—C_2H_5$

*Most common monosaccharides, for example, glucose, are polyhydroxy alcohols.

†Dicarboxylic and tricarboxylic acids (containing two and three COOH groups respectively) occur commonly in cells. Hydroxy acids (containing additional hydroxyl groups) and keto acids (containing additional carbonyl groups) are also organic acid variants frequently found in plants and animals.

‡Amino acids are the building blocks of proteins.

§The characteristic tastes and odors of certain fruits (for example, bananas, pineapples, apricots) are due to the presence of esters.

tones, carboxylic acids, amino acids, ethers, and *esters.* In each case, the characteristic groups added to an alkyl group (R—) determine the class of compound that is formed. Many of the hydrocarbon derivatives are constituents of living cells and tissues. Table 4-3 summarizes the main features of these compounds.

RING COMPOUNDS

Carbon compounds occur in ring as well as chain forms. Chain hydrocarbons, such as pentane (C_5H_{12}) and hexane (C_6H_{14}), can form *cyclic* compounds, such as cyclopentane (C_5H_{10}) and cyclohexane (C_6H_{12}), when the terminal carbons are bonded to each other instead of to 2 hydrogen atoms:

cyclic

```
    H  H  H  H  H                 CH2
    |  |  |  |  |                /   \
 H—C—C—C—C—C—H             H2C     CH2
    |  |  |  |  |              |       |
    H  H  H  H  H              H2C———CH2
   normal pentane               cyclopentane

    H  H  H  H  H  H             H2C—CH2
    |  |  |  |  |  |            /       \
 H—C—C—C—C—C—C—H         H2C         CH2
    |  |  |  |  |  |            \       /
    H  H  H  H  H  H             H2C—CH2
     normal hexane               cyclohexane
```

The molecular formulas differ slightly for the normal and cyclic hydrocarbons, as do the properties of the different molecules. Note also that the cyclopentane molecule is a five-membered ring (pentagon), whereas the cyclohexane is in the form of a six-membered ring (hexagon).

A most important ring compound of carbon is *benzene* (C_6H_6). This is a 6-carbon ring with the following structural formulas:

```
        H                          H
        |                          |
        C                          C
      //  \                      /  \\
  H—C      C—H              H—C      C—H
     |      ||      and         ||      |
  H—C      C—H              H—C      C—H
      \\  /                      \  //
        C                          C
        |                          |
        H          benzene         H
```

benzene ring

The *benzene ring* is a particularly stable compound. It is the parent compound for many important substances, often called *aromatic* hydrocarbons, which are formed by substituting different groups for the hydrogen atoms on the benzene ring. In the structural formulas for the benzene ring shown above, the alternating single and double bonds between the carbons around the ring are shown occupying different positions. Actually, neither formula is a correct version of the benzene ring; the single and double pairs of shared electrons are at all times oscillating between the two positions shown above. This phenomenon is called *resonance;* the true structure of the benzene ring is said to be a *resonance hybrid,* with the six carbon-to-carbon bonds neither completely single nor completely double.

resonance

A variety of compounds formed either by replacement of 1 or more hydrogen atoms on the benzene ring or by the combination of two or more benzene rings (which mutually share carbon atoms) are listed in Table 4-4. For convenience in writing the structural formula of benzene and its derivatives, the compound is usually depicted simply as a hexagon, with the understanding that there is a —CH group at each corner.

Heterocyclic Compounds

heterocyclic

When 1 or more of the carbon atoms in cyclic compounds, or in the benzene ring, are replaced by some other element, *heterocyclic compounds* are formed. A group of five- and six-membered heterocyclic ring compounds of great biological importance are those in which nitrogen atoms replace 1 or 2 of the carbon atoms.

1. Imidazole: a five-membered ring compound that is a component of the amino acid histidine and also forms part of the purines found in nucleic acids.

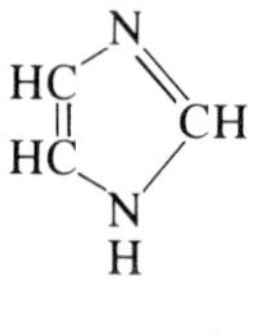

N
HC
CH
HC
N
H

N
N
H

structural formula

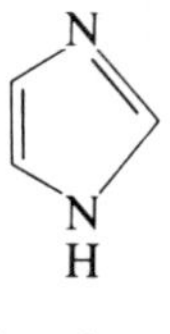

abbreviated structural formula

2. Pyrimidine: a six-membered ring. Derivatives form the heterocyclic bases cytosine, uracil, and thymine, the building blocks of the nucleic acids.

Table 4-4. Typical compounds derived from benzene (C_6H_6)

Name	Molecular formula	Abbreviated structural formula
By substitution of 1 or more hydrogen atoms:		
Parent compound:		
Benzene	C_6H_6	
Toluene	$C_6H_5CH_3$	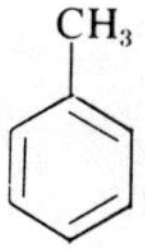 CH_3
Phenol	C_6H_5OH	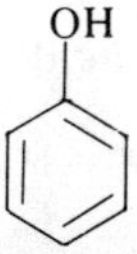 OH
Aniline	$C_6H_5NH_2$	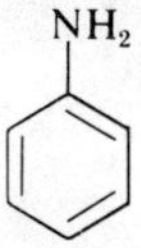 NH_2
Benzoic acid	C_6H_5COOH	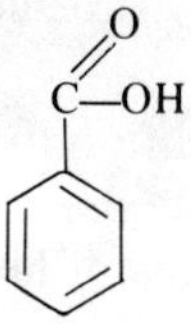 O, C—OH
Paradichlorobenzene	$C_6H_4Cl_2$	 Cl, Cl
By condensation of 2 or more benzene rings:		
Naphthalene	$C_{10}H_8$	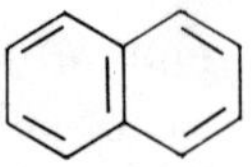
Anthracene	$C_{14}H_{10}$	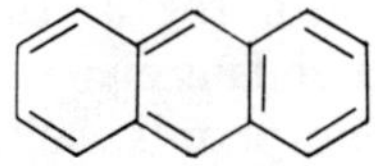
Phenanthrene* (an isomer of anthracene)	$C_{14}H_{10}$	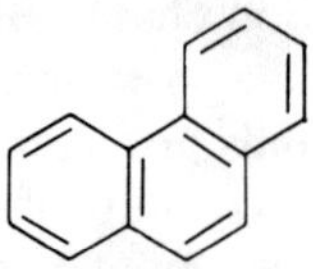

*The steroid nucleus, perhydrocyclopentanophenanthrene, which contains a cyclopentane ring in addition to the phenanthrene rings, is the parent compound for cholesterol, the sex hormones, and other important substances found in the body (see Chapter 6):

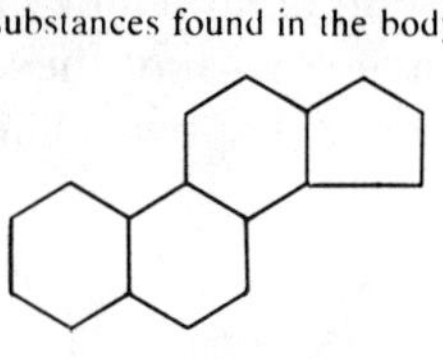

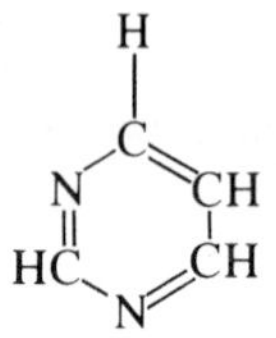

structural formula 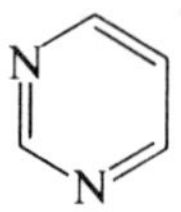 | abbreviated structural formula

3. Purines: compounds composed of pyrimidine and imidazole rings fused together. Derivatives are building blocks of nucleic acids, for example, adenine and guanine.

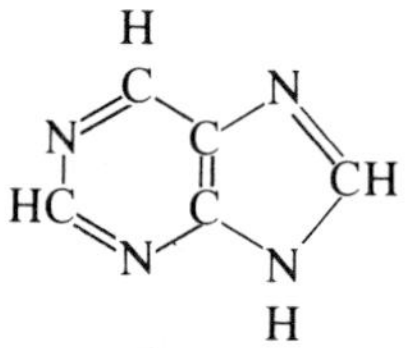

structural formula 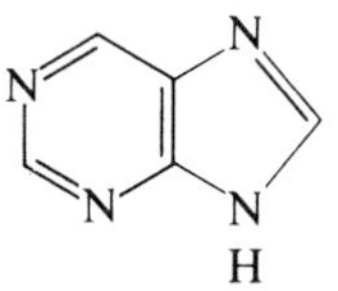 | abbreviated structural formula

In addition to their supreme importance as components of the nucleic acids, which are the chemical basis of heredity, many derivatives of heterocyclic nitrogen compounds have pharmacological activity and are in common use as drugs. These include substances widely distributed in the plant kingdom, such as caffeine, nicotine, and the alkaloids of belladonna and opium. Other derivatives, such as the barbiturates, are manufactured.

CARBON COMPOUNDS AND BIOLOGY

The bulk of the solid material in cells consist of organic compounds made of carbon in combination with hydrogen, oxygen, and nitrogen. Together these four elements constitute the most abundant elements in living organisms. Other elements such as phosphorus, sulfur, metals, and halogens are also found on the carbon chains and rings of biological organic molecules.

The possible variations that can be introduced into carbon compounds are almost endless. Chains and rings may vary in length and size; different functional groups may be added to carbon backbones; bonds may be single or multiple; isomerism (different three-dimensional arrangements) adds even more diversity to organic molecules. Carbon compounds can undergo many diverse types of chemical change. Most of these are not within the scope of this discussion. However, two characteristic chemical reac-

Table 4-5. Major classes of biological polymers and building blocks

Type of compound	Building blocks
Polysaccharides (*carbohydrates* such as starch and glycogen)	Monosaccharides (mainly 6-carbon sugars, or hexoses)
Triacylglycerols (neutral *fats*)	Fatty acids (hydrocarbons of varying chain lengths) Glycerol
Polypeptides (one or more polypeptide chains constitute a *protein*)	Amino acids
Polynucleotides (single or double polynucleotide chains constitute the *nucleic acids*, DNA, and RNA)	Mononucleotides (combination of a purine or pyrimidine, 5-carbon monosaccharide, and a phosphate group)

tions will be mentioned here because they are the basis of much of the chemistry of living processes.

Polymerization Reactions

polymer

The term *polymer* refers to very large molecules formed by the repeated combination of smaller units (monomers). *Polymerization* is a distinctive reaction of carbon compounds. The four major divisions of biological organic substances—carbohydrates, lipids, proteins, and nucleic acids—are polymers.* Since they are found only in living matter, these compounds and their derivatives have been aptly termed *biomolecules* (*bios,* life). Many of them are giant molecules with masses of 10^6 daltons (d) or more; hence, the term *macromolecule* (*macro,* large). Although the four categories consist of extremely diversified molecules, they have one feature in common: they are all composed of specific small units, or *building blocks,* covalently linked together in chains of varying lengths (Table 4-5). Each linked unit in the polymer is conventionally termed a *residue*. When living cells engage in the process of *biosynthesis,* that is, when they assemble these large polymers, they do so in a manner analogous to building a brick wall by cementing

residue
biosynthesis

* A special branch of chemistry (polymer chemistry) is concerned with the manufacture of artificial polymers, for example, fibers and plastics such as nylon, Orlon, polyesters, polyvinyl, or polyurethane. In many respects, these man-made products mimic the properties of naturally occurring polymers, for example, silk and wool (proteins); cellulose derivatives such as wood, cotton, or linen (carbohydrates); and rubber (plant hydrocarbons).

dehydration synthesis

many individual bricks, or building blocks, together. This particular type of polymerization reaction is called *dehydration synthesis* (or condensation); the word *dehydration* refers to the fact that one small unit is connected to the next by the removal of a molecule of water. In such reactions, each building block molecule contributes either a hydrogen atom (H) or a hydroxyl group (OH) to the formation of water (H_2O). At the same time, the adjacent ends of the monomers are joined together. By repetition of this process, large complex polymers are formed.

Hydrolysis

hydrolysis

The breakdown (decomposition) of large molecules into smaller molecules, with *water as a reactant,* is called *hydrolysis*. Hydrolysis reactions are the reverse of dehydration synthesis. In the hydrolytic breakage of the covalent bonds that connect the building block units, molecules of water replace the missing hydrogen and hydroxyl groups that were originally removed when the monomers were linked together. Hydrolysis is the most common type of reaction for the breakdown of large molecules into smaller ones, the characteristic form of decomposition reaction in living organisms.

A diagram illustrating the basic design of dehydration synthesis and hydrolysis as carried out by living organisms is given in Fig. 4-4. The properties of organic compounds of biological origin are discussed in the chapters that follow. As we shall see, the building block units of biological polymers subserve *more than one* func-

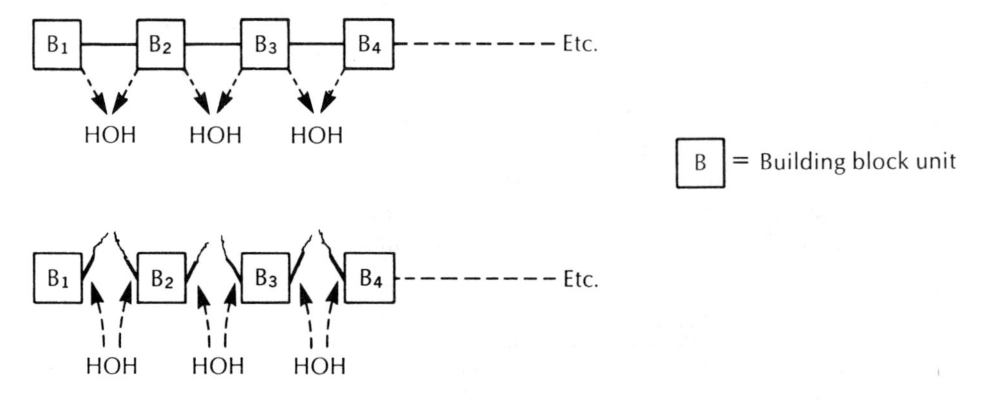

Fig. 4-4. Schematic model of polymerization and breakdown reactions in cells.

tion in cells; that is, in addition to providing the raw material for the synthesis of macromolecules, they also act as *precursors* (starter compounds) for a variety of other physiologically active substances, and they are vitally necessary fuels for the production of energy.

precursors

5

CARBOHYDRATES

The term *carbohydrate* refers to sugars and starches of plant and animal origin. The most abundant sources of these food molecules on earth are green plants, which produce carbohydrates through the process of photosynthesis:

$$CO_2 + H_2O + \text{sunlight} \xrightarrow{\text{chlorophyll}} \text{carbohydrates} + O_2$$

Many of these compounds, composed of the elements carbon, hydrogen, and oxygen, have the empirical formula $(CH_2O)_n$ (n may be a small or a very large number), in which the ratio of hydrogen to oxygen is the same as that in water, namely 2:1. However, there are carbohydrates that do not have this H:O ratio, as well as carbohydrate derivatives that contain nitrogen, phosphorus, and sulfur, in addition to carbon, hydrogen, and oxygen. In living organisms, carbohydrates are the most important source of fuel for cell metabolism and are also major functional and structural constituents of cells and tissues.

The main types of carbohydrates occur in a variety of molecular sizes, ranging from the smallest units—the simplest sugars, or *monosaccharides*—to large complex molecules called *polysaccharides*. The polysaccharides are composed of many simple monosaccharide units linked together in long chains. This is a characteristic pattern in cellular chemistry and applies generally to many biological molecules: small units, or *monomers* (single parts), are the building blocks for large molecules, or *polymers* (many parts). The cells of the body possess the necessary chemical mechanisms for the biosynthesis of the large polymeric molecules, or macromolecules, from the small building block units and for the con-

monomers
polymers

verse process, the breakdown of the polymers into their component monomers.

Monosaccharides

monosaccharides

Monosaccharides are defined as simple sugars that cannot be further broken down by hydrolysis (see Chapter 4) to simpler carbohydrates. They are mainly polyhydroxy aldehydes or ketones, containing from 3 to 7 carbon atoms. The group names of monosaccharides are derived from the numbers of carbon atoms in the chains:

3-carbon sugars—trioses

4-carbon sugars—tetroses

5-carbon sugars—pentoses

6-carbon sugars—hexoses

7-carbon sugars—heptoses

In the body, the hexoses are the most commonly found monosaccharides. Two pentoses, ribose and deoxyribose, are important constituents of the nucleic acids and the energy storage and regulator compounds of cells (see chapter 11). A triose, glyceraldehyde, is a major intermediate in cellular carbohydrate metabolism (see Chapter 10).

The three hexoses of biological importance—glucose, fructose, and galactose—all have the chemical formula $C_6H_{12}O_6$. Their open-chain structures can be represented as shown in Fig. 5-1.

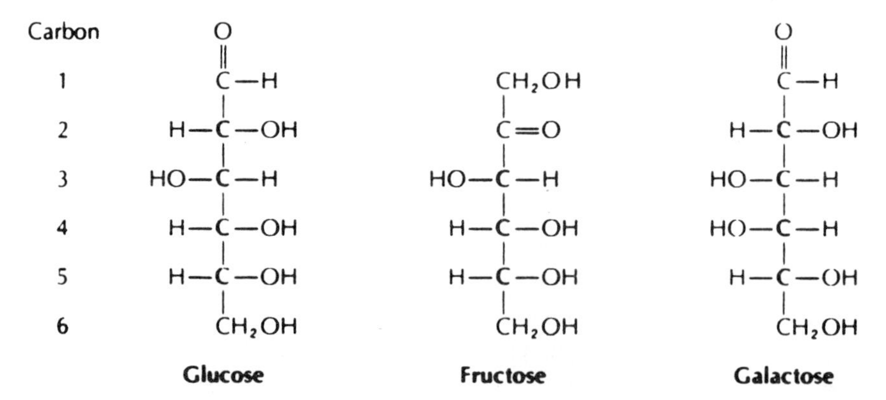

Fig. 5-1. Structural formulas for common monosaccharides. Column at left indicates usual system of numbering the carbon atoms of sugars. Carbons shown in darker print are asymmetrical—bonded to four different groups.

The carbon atoms are often numbered as indicated in the left column of the figure.

If the carbonyl group (C═O) is on carbon 1, the monosaccharide is an aldehyde and is called an *aldose;* if it is on an internal carbon, the monosaccharide is classed as a ketone and is called a *ketose*. As can be seen from the structural formulas of the hexoses, glucose and galactose are aldoses and fructose is a ketose.

aldose

ketose

The aldehyde and keto groups (carbonyls) are called *reducing groups* because they will chemically reduce certain metal ions (for example, copper). This property is utilized in the laboratory analysis of sugars as found in urine or blood. The reduced form of copper sulfate is red and insoluble.

reducing groups

Isomerism. Glucose, fructose, and galactose are *structural isomers* of each other. They have the same chemical formula ($C_6H_{12}O_6$) but have different structural formulas and are distinct chemical compounds. In fact, these sugars are three-dimensional isomers because they differ only by the orientation of groups around their asymmetrical carbon atoms. Each of the internal carbons (2 through 5 for glucose and galactose, 3 through 5 for fructose) is bonded to four different groups: an H, an OH, and two other larger groups that amount to the rest of the molecule. Therefore, it makes a difference how they are arranged (as described in Chapter 4).

structural isomers

By convention, the open-chain structural formulas are drawn as in Fig. 5-1, with the carbon chain arranged vertically and the hydrogen and hydroxyl groups either to the left or right. In the ball-and-stick model of Chapter 4, none of the bonds would lie in the plane of this page. If a given C is considered as in that plane, then the vertical bonds would be pointing back at an angle, and the H and OH bonds would be pointing toward the reader at the same angle, as in Fig. 5-2. So, in two dimensions, the mirror images of a particular asymmetrical carbon atom are indicated as

$$\text{H—}\overset{|}{\underset{|}{\text{C}}}\text{—OH} \quad \text{and} \quad \text{HO—}\overset{|}{\underset{|}{\text{C}}}\text{—H}$$

Glucose and galactose differ only by the orientation of groups around carbon 4. Glucose and fructose differ in that carbon 2 of fructose is not asymmetrical; rather, it is the carbonyl carbon (with only three different groups attached).

Each of these monosaccharides has a mirror image. The two forms of glucose are shown in Fig. 5-3. When glucose is synthesized in the laboratory, the product is a 50-50 mixture of the mirror

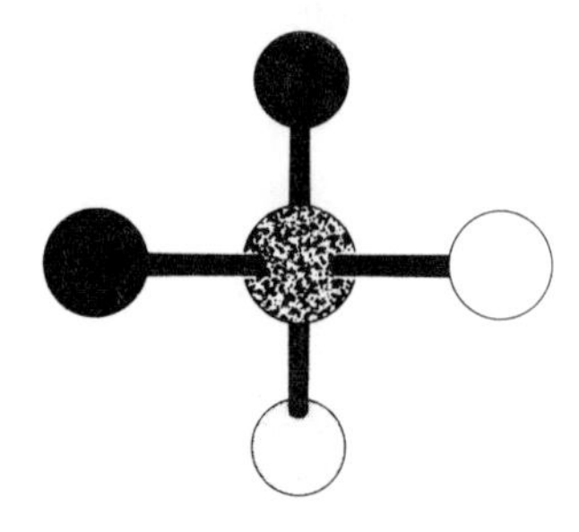

Fig. 5-2. A three-dimensional representation of an asymmetrical carbon atom (center), oriented as in the conventionally drawn open-chain structure formula of a monosaccharide. The black and white balls represent H and OH groups; the gray balls are the upper and lower portions of the rest of the sugar molecule. The H and OH groups should be regarded as projecting forward and the other groups back from the plane of the paper.

images. But if glucose is produced in a biological system, it is 100% D-glucose (D = dextro, *right*). Living systems neither produce nor utilize the mirror image L-glucose (L = levo, *left*), because of their dependence on enzymes (see Chapter 8), which are three-dimensionally discriminating. In fact, all the biologically active monosaccharides have this orientation

H—C—OH

D-glucose	L-glucose
O=C—H	O=C—H
H—C—OH	HO—C—H
HO—C—H	H—C—OH
H—C—OH	HO—C—H
H—C—OH	HO—C—H
CH_2OH	CH_2OH

Fig. 5-3. Structural formulas for the mirror images D- and L-glucose. Note that the arrangements of groups around carbons 1 and 6 are shown as the same in both forms. These carbons are symmetrical and are therefore equivalent, whichever way they may be rotated. Carbons 2 to 5 are asymmetrical and possess optical activity.

around the carbon atom farthest from the reducing group. In hexoses, this is the number-5 carbon. So the hydroxyl group on carbon 5 is always to the right in D-sugars.

Asymmetrical carbon atoms possess a property not shared by symmetrical forms. That property is *optical activity*—the rotation of plane-polarized light. If a plane of polarized light passes through a solution of a compound that has an asymmetrical carbon, then that plane of light will be rotated to the right (+) or to the left (−). It is seldom possible to predict the rotation from the formula of a given compound, but it is known that whatever the rotation by one isomer, its mirror image will rotate light exactly the same degree but in the opposite direction. For example, if a certain concentration of D-glucose has a rotation of +52°, then the same concentration of L-glucose will rotate the light −52°. A 50-50 mixture of D-glucose and L-glucose has no optical activity. It is important to note that the designation D or L has nothing to do with the *direction* of rotation; rotation is dependent on the organized structure of whatever asymmetrical atoms are in the molecule. D and L refer to the three-dimensional configuration around that *one* asymmetrical carbon atom, the one farthest from the reducing group in the sugar. The optically active isomeric forms of any particular compound are called *enantiomers* (*enantio,* opposite; *meros,* part).

optical activity

enantiomers

Ring structures of the monosaccharides. Only very small amounts of the monosaccharides are found in the open chain form. Most of the molecules are five- or six-membered heterocyclic ring structures. The rings result from intramolecular reactions between the very active carbonyl (C=O) group and the hydroxyl group on the next to the last carbon atom. These groups are brought into close proximity because of the angles of the carbon-carbon bonds. Glucose and galactose are found as six-membered (pyranose) rings, whereas fructose usually forms a five-membered (furanose) ring. The pentoses ribose and deoxyribose also occur in the body as five-membered ring structures (see Chapter 11). Pentose and hexose ring structures are usually depicted in an abbreviated form without the carbons and hydrogens and with lines representing the —OH groups. In the ring structures of glucose and fructose shown in Fig. 5-4, most of the carbon atoms are numbered to show the internal arrangements of the molecules.

Derivatives of monosaccharides. A number of biologically important compounds found in living organisms are derivatives of monosaccharides in which one or more of the groups attached to the carbon atoms are changed.

Phosphoric acid esters of hexoses and pentoses are present in all cells. The 5-carbon sugars ribose and deoxyribose are invaria-

D-glucopyranose D-fructofuranose

Fig. 5-4. Simplified structural formulas for commonly existing cyclic forms of glucose (D-glucopyranose) and fructose (D-fructofuranose). The positions of most of the carbons are numbered outside the rings.

bly linked with phosphoric acid groups in active biomolecules such as the nucleotides adenosine monophosphate (AMP), adenosine diphosphate (ADP), and adenosine triphosphate (ATP), and the polynucleotides DNA and RNA. The hexoses glucose and fructose enter into very few biological reactions without first being converted to phosphoric acid esters. This process of *phosphorylation* is catalyzed by cell enzymes.

phosphorylation

amino sugars

Two *amino sugars* found in combined forms in many tissues of the body are glucosamine and galactosamine. The hydroxyl (OH) group of the carbon atom 2 in these sugars is replaced by an amino (NH_2) group, which is usually present in a substituted NH— form.

sugar acid

When the hydroxyl group at carbon atom 6 in glucose is oxidized to a carboxyl (—COOH) group, a *sugar acid,* glucuronic acid, is formed. This acid is an important constituent of glycosaminoglycans found in the body, and is used by liver cells as a coupling agent. Many drugs, potentially toxic substances such as phenol and benzoic acid, and a number of steroid hormones, for example, estrogens, are solubilized and inactivated by being conjugated with glucuronic acid in the liver. They are then excreted in the urine as glucuronides.

ascorbic acid

Another important 6-carbon sugar acid is *ascorbic acid,* or vitamin C (Fig. 5-5). It can be seen from the structural formula that ascorbic acid lacks the characteristic carboxyl group (COOH) of organic acids; its acidity is due to the interchange (resonance) of the single and double bonds on the first 3 carbon atoms. Ascorbic acid is water soluble and optically active (vitamin C is L-ascorbic acid). Humans and other primates, in common with insects, fish, bats, and guinea pigs, are unable to synthesize this vitally necessary sugar acid from hexose precursors and are thus susceptible to the deficiency disease, scurvy, when their dietary intake of

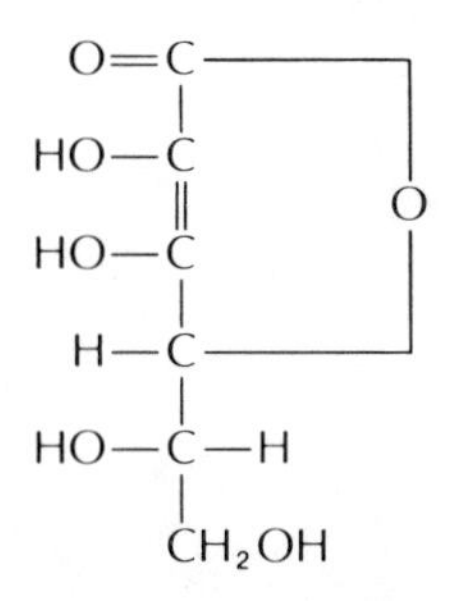

Ascorbic acid (vitamin C)

Fig. 5-5. Structural formula of ascorbic acid (vitamin C).

vitamin C is low. Ascorbic acid is mainly required by the body for the biosynthesis of collagen, a major structural protein of fibrous connective tissues, bone, and cartilage. The symptoms of avitaminosis C (deficiency of vitamin C) are due in large measure to impaired collagen formation. Citrus fruits and fresh vegetables provide good dietary sources of the vitamin.

A more complex group of sugar acids that additionally contain substituted NH— groups are the *sialic acids*. Sialic acids, such as N-acetylneuraminic acid, and other amino sugars are commonly found in the oligosaccharide chains of glycolipids and glycoproteins (see p. 95).

Disaccharides

disaccharides

The *disaccharides* consist of two monosaccharide units linked together by a reaction in which 1 molecule of water (H_2O) is given off. A diagram of the condensation of two monosaccharides to form a disaccharide is shown in Fig. 5-6. The three important disaccharides are maltose, lactose, and sucrose. They all have the same molecular formula ($C_{12}H_{22}O_{11}$) and are thus structural isomers of each others.

hydrolysis

Maltose is formed as an intermediate product in the breakdown of starch to glucose. It consists of 2 molecules of glucose condensed together. In the body, the linkage between the two glucose residues is broken by the process of *hydrolysis* (a decomposition reaction in which water is one of the reactants):

$$\text{maltose} + H_2O \xrightarrow{\text{hydrolysis}} 2\ \text{glucose}$$

$$(C_{12}H_{22}O_{11} + H_2O \rightarrow 2\ C_6H_{12}O_6)$$

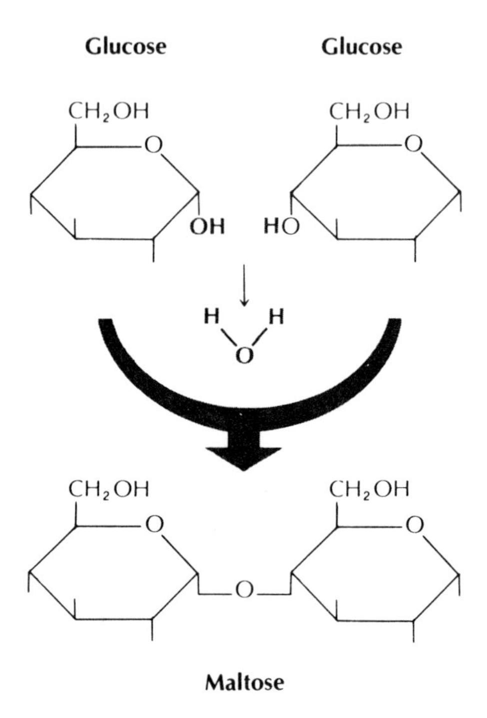

Fig. 5-6. Synthesis of the disaccharide maltose from 2 glucose molecules.

Lactose, or milk sugar, is a disaccharide found in milk and is synthesized by the mammary gland. On hydrolysis, it yields the monosaccharides glucose and galactose:

$$\text{lactose} + H_2O \xrightarrow{\text{hydrolysis}} \text{glucose} + \text{galactose}$$

Sucrose, or cane sugar, occurs abundantly in plants and is an important dietary carbohydrate. On hydrolysis, it yields the monosaccharides glucose and fructose:

$$\text{sucrose} + H_2O \xrightarrow{\text{hydrolysis}} \text{glucose} + \text{fructose}$$

Polysaccharides

polysaccharides

Polysaccharides are formed by the condensation of large numbers of monosaccharide or monosaccharide-derivative units. The three main plant and animal polysaccharides—starch, cellulose, and glycogen—are built up entirely from glucose units.

Starch is a high molecular weight storage carbohydrate of plant cells. It is readily digested by man, first by partial hydrolysis to maltose (disaccharide) and then finally to glucose (monosaccharide), by the action of the enzymes amylase and maltase (see Chapter 8).

Cellulose, which is probably the most abundant organic chemical on earth, is a structural carbohydrate that forms plant cell walls and other supporting tissues of plants, for example, plant fibers and wood. The linkages between the glucose units in cellulose differentiate it from starch and glycogen and are responsible for the fact that it is resistant to the amylases present in the human digestive tract. It is thus not digested to any extent by man. However, certain bacteria and protozoa contain enzymes that can hydrolyze cellulose, and their presence in the rumen or colon of herbivorous animals enables these animals to utilize cellulose as a food molecule.

Glycogen is the storage carbohydrate compound of animal tissues, particularly liver and skeletal muscle tissue. It is a branched polymer of thousands of glucose units, with a minimum molecular weight of 5,000,000 (Fig. 5-7). Liver cells can convert other hexoses to glucose by isomeric rearrangement, and then glucose

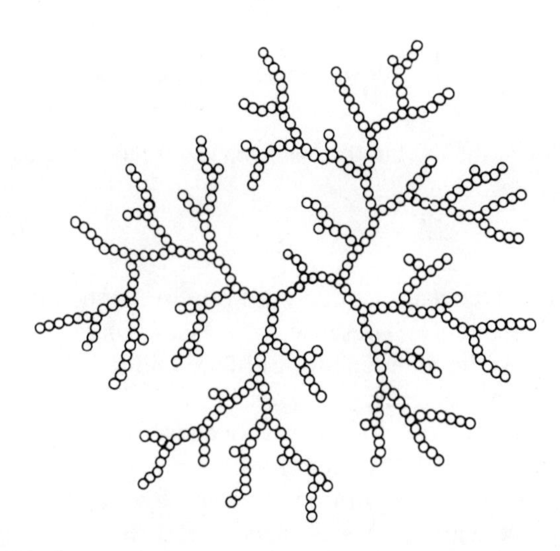

Fig. 5-7. Diagrammatic version of the branched structure of glycogen. Each small circle represents 1 glucose molecule. Most linkages are between carbon 1 of one glucose unit and carbon 4 of the next (1,4 linkages). Some glucose units, however, have chains attached to their carbon 6 as well as carbon 4. The combinations of these 1,6 linkages with the 1,4 linkages result in the branching. (From Hickman, C. P., and others: Integrated principles of zoology, ed. 5, St. Louis, 1974, The C. V. Mosby Co.)

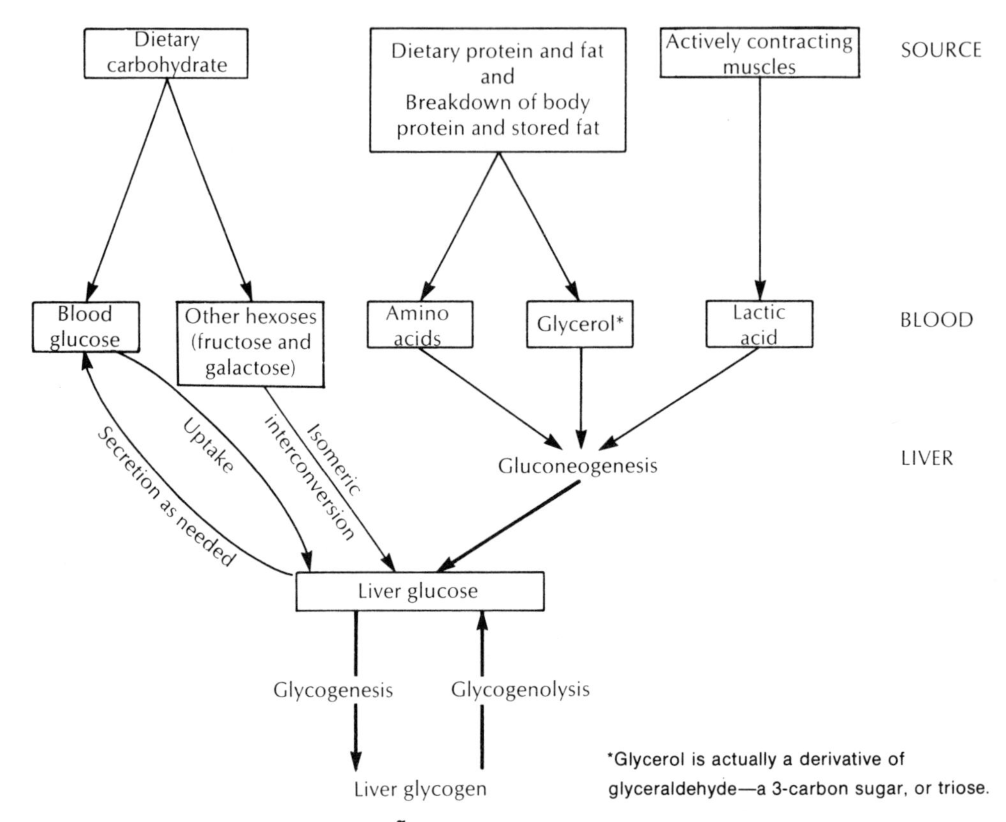

Fig. 5-8. Role of the liver in the maintenance of blood glucose homeostasis.

glycogenesis
glycogenolysis

gluconeogenesis

to glycogen. Glycogen is stored until required, when it is broken down again to glucose. These two processes, which are essential for the homeostatic maintenance of blood and tissue glucose levels, are called respectively *glycogenesis* and *glycogenolysis*. The main source of glucose for the glycogenesis that takes place in the liver is dietary carbohydrate. However, in the absence of adequate amounts of carbohydrate in the diet, liver cells can convert noncarbohydrate precursors to glucose and thence to glycogen. This process is called *gluconeogenesis*. The conversions can involve lactic acid (formed by skeletal muscles during exercise), glycerol (an end product of fat hydrolysis), and some amino acids. Since glucose is the major fuel used by living cells as a source of energy, survival is dependent on the maintenance of glucose levels in the body within relatively narrow physiological limits (normal values in blood are 65 to 100 mg/dL). The liver plays a central role in the complex feedback mechanisms that regulate glucose homeostasis (Fig. 5-8).

Carbohydrate/Protein and Carbohydrate/Lipid Complexes

Proteins and lipids combined covalently with diverse sugars constitute an important group of structural and functional biomolecules. *Glycoproteins* (*glykys,* sweet) are proteins with oligosaccharide chains attached to them. The oligosaccharides are typically composed of 10 or fewer sugar residues (*oligo,* few). As well as being major components of cell membranes, examples of glycoproteins include, among others, collagen (a structural protein widely distributed in the body), numerous enzymes and hormones, mucin (mucus), antibodies, and the blood group antigens.

glycoproteins

The carbohydrate-containing *glycolipids* are fatty compounds, such as *cerebrosides* and *gangliosides,* which are also found in cell membranes. In addition to their lipid moieties, the cerebrosides contain either glucose or galactose. The gangliosides have more complex oligosaccharide chains, often with sialic acid residues.

glycolipids

The *proteoglycans* (formerly known as mucopolysaccharides) are a group of macromolecular complexes that are composed of about 95% polysaccharide and 5% protein. The polysaccharide components, termed *glycosaminoglycans,* are in the form of long linear chains of repeating disaccharide units consisting of an amino sugar and a sugar (uronic) acid often esterified with sulfate groups. In most cases, the linear chains are attached to a core protein. Due to the presence of the sugar acids and sulfate groups, these compounds are strongly negatively charged, with acidic properties. They tend to form highly viscous, lubricating and noncompressible solutions in water. The proteoglycans are found in the intercellular ground substances of connective tissues (i.e., cartilage, tendons, ligaments, etc.). Other examples are *hyaluronic acid,* a constituent of the vitreous humor of the eye and of synovial (joint) fluid, and *heparin,* an anticoagulant substance produced by connective tissue mast cells and basophils.

proteoglycans

6

LIPIDS

Lipids are a diverse group of fatty substances that vary in chemical structure but have one feature in common: they are to a large extent insoluble in water, but soluble in organic solvents such as chloroform, carbon tetrachloride, ether, and benzene; some lipids are also soluble in alcohol and acetone.

The following physiologically important lipids and lipid derivatives will be discussed in this chapter:

1. Fatty acids
2. Triacylglycerols (also called neutral fats or triglycerides)
3. Phospholipids and glycolipids
4. Steroids, i.e., cholesterol and its derivatives the steroid hormones (sex hormones and hormones of the adrenal cortex), bile acids, and vitamin D (a fat-soluble vitamin)
5. Other fat-soluble vitamins: vitamins A, E, and K
6. Prostaglandins
7. Serum lipoproteins

FATTY ACIDS

fatty acids

The *fatty acids* subserve important functional and structural purposes in the body. They are the building blocks of triacylglycerols, the form in which fat is stored in the body, and a reserve of energy-rich metabolic fuel. They are also the building blocks (or precursors) of other vital lipid substances, such as the phospholipid and glycolipid components of cellular membranes, and the prostaglandins.

Fatty acids are straight-chain *aliphatic hydrocarbon* organic acids, usually containing one carboxyl group. Their general formula is

$$CH_3(CH_2)_nCOOH$$

saturated

unsaturated

polyunsaturated

essential fatty acids

The value of n in naturally occurring fats is between 10 and 20, and n is always an even number. Fatty acids that have only single bonds between the carbon atoms of the chain are said to be *saturated* (there is no room for additional atoms in the molecule). *Unsaturated* fatty acids have one or more double bonds between the carbon atoms of the chain. If they have more than one double bond, they are *polyunsaturated* (see Table 6-1). The polyunsaturated fatty acids, linoleic acid and linolenic acid, are considered to be *essential fatty acids* in man. The term 'essential' implies that these acids must be supplied by dietary fats because they cannot be synthesized in the body. Using linoleic and linolenic acids as precursors, liver cells *(hepatocytes)* can synthesize arachidonic acid and other necessary polyunsaturated fatty acids. Structural formulas of some short and longer chain fatty acids are shown in Fig. 6-1.

The melting points of unsaturated fatty acids are lower than those of saturated fatty acids, so that fats containing unsaturated fatty acids are usually liquid at room temperature. The latter are commonly called *oils*. Most fats of vegetable (plant) origin, which generally have a preponderance of polyunsaturated fatty acids, are oils, for example, olive oil, peanut oil and sunflower seed oil. The exceptions here are palm oil and coconut oil; these vegetable oils contain saturated fatty acids. Saturated fatty acids likewise

Table 6-1. Some naturally occurring fatty acids

	Number of carbon atoms	Number of double bonds	Chemical formula
Saturated fatty acids			
Lauric acid	12		$C_{11}H_{23}COOH$
Palmitic acid	16		$C_{13}H_{31}COOH$
Stearic acid	18		$C_{17}H_{35}COOH$
Arachidic acid	20		$C_{19}H_{39}COOH$
Unsaturated fatty acids			
Palmitoleic acid	16	1	$C_{13}H_{29}COOH$
Oleic acid	18	1	$C_{17}H_{33}COOH$
Linoleic acid*	18	2	$C_{17}H_{31}COOH$
Linolenic acid*	18	3	$C_{17}H_{29}COOH$
Arachidonic acid*	20	4	$C_{19}H_{31}COOH$

* These three fatty acids are polyunsaturated.

```
      H   H   H   O
      |   |   |   ||
  H—C—C—C—C—OH
      |   |   |
      H   H   H
```

Butyric acid (C_3H_7COOH), found in butter

```
    H  H  H  H  H  H  H  H  H  H  H  O
    |  |  |  |  |  |  |  |  |  |  |  ||
 H—C—C—C—C—C—C—C—C—C—C—C—C—OH
    |  |  |  |  |  |  |  |  |  |  |
    H  H  H  H  H  H  H  H  H  H  H
```

A saturated fatty acid, lauric acid ($C_{11}H_{23}COOH$), found in coconut oil

```
    H  H  H  H  H  H  H  H  H  H  H  H  H  H  H  H  H  O
    |  |  |  |  |  |  |  |  |  |  |  |  |  |  |  |  |  ||
 H—C—C—C—C—C—C—C—C—C=C—C—C—C—C—C—C—C—C—OH
    |  |  |  |  |  |  |  |        |  |  |  |  |  |  |
    H  H  H  H  H  H  H  H        H  H  H  H  H  H  H
```

An unsaturated fatty acid, oleic acid ($C_{17}H_{33}COOH$), found in olive oil

Fig. 6-1. Some common fatty acids.

predominate in fats from animal sources (including human body fat). However, fats obtained from fish are largely unsaturated. It may be noted in this context that margarine is made by partially hydrogenating vegetable oils. The addition of hydrogen solidifies the oils, forming a spreadable product, but it also saturates some of the unsaturated fatty acids in the oils. Diets rich in saturated fats are now recognized to be one of the factors associated with an increased risk of heart attacks.

TRIACYLGLYCEROLS

triacylglycerols

Fatty acids are stored in the adipose tissue of the body in the form of *triacylglycerols*. Large droplets of triacylglycerols are characteristic inclusions in the cytoplasm of the cells of this tissue (*adipocytes* or *fat cells*). This concentrated stored lipid serves as a reservoir of high-calorie metabolic fuel which is mobilized as needed when the body requires a source of energy. Older terms for triacylglycerols are triglycerides or neutral fats.

The building blocks of triacylglycerols are fatty acids and glycerol. The compound *glycerol* is a trihydric alcohol (containing

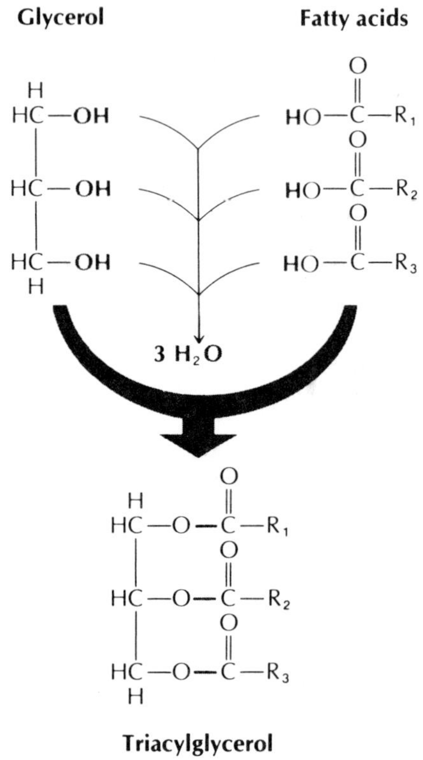

Glycerol　　　Fatty acids

$$\begin{array}{ll} H & \\ HC-OH & HO-\overset{O}{\overset{\|}{C}}-R_1 \\ | & \\ HC-OH & HO-\overset{O}{\overset{\|}{C}}-R_2 \\ | & \\ HC-OH & HO-\overset{O}{\overset{\|}{C}}-R_3 \\ H & \end{array}$$

$3\ H_2O$

$$\begin{array}{l} H \\ HC-O-\overset{O}{\overset{\|}{C}}-R_1 \\ | \\ HC-O-\overset{O}{\overset{\|}{C}}-R_2 \\ | \\ HC-O-\overset{O}{\overset{\|}{C}}-R_3 \\ H \end{array}$$

Triacylglycerol

Fig. 6-2. Formation of a triacylglycerol.

three hydroxyl groups) with the chemical formula $C_3H_5(OH)_3$. Its structural formula is as follows:

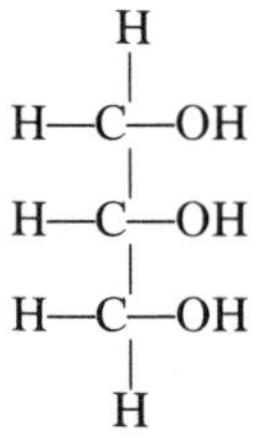

$$\begin{array}{c} H \\ | \\ H-C-OH \\ | \\ H-C-OH \\ | \\ H-C-OH \\ | \\ H \end{array}$$

In the formation of a triacylglycerol, three fatty acids form esters with the three hydroxyl groups of glycerol. As shown in Fig. 6-2, three molecules of water are given off as the ester bonds are formed. Each water molecule consists of a hydrogen from the carboxyl (—COOH) group of one of the fatty acids, and a hydroxyl group (—OH) from the glycerol molecule. If only one fatty acid esterifies to the glycerol molecule, the product is called a monoacylglycerol (or monoglyceride); if two fatty acids bind to glycerol, a diacylglycerol (or diglyceride) is formed.

The structure of a typical triacylglycerol may be shown as follows (the R groups represent the hydrocarbon chains of the fatty acids):

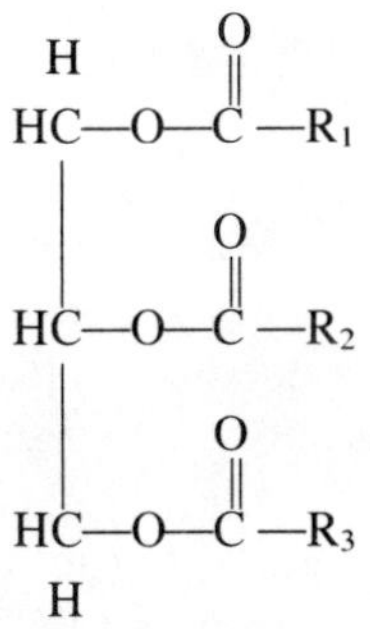

$$
\begin{array}{l}
\quad\;\; \mathrm{H} \qquad\;\; \mathrm{O} \\
\quad\qquad\qquad\;\; \| \\
\mathrm{HC{-}O{-}C{-}R_1} \\
\quad | \qquad\quad\;\; \mathrm{O} \\
\quad | \qquad\quad\;\; \| \\
\mathrm{HC{-}O{-}C{-}R_2} \\
\quad | \qquad\quad\;\; \mathrm{O} \\
\quad | \qquad\quad\;\; \| \\
\mathrm{HC{-}O{-}C{-}R_3} \\
\quad\;\; \mathrm{H}
\end{array}
$$

When R_1, R_2, and R_3 are the same fatty acid, such as in triolein, a triacylglycerol found in olive oil (three oleic acids plus glycerol), the compound is called a *simple triacylglycerol.* Compounds formed by two or more different fatty acids are *mixed triacylglycerols*. In the formation of these esters, it can be seen that the presence of three hydroxyl groups in the glycerol molecule allows for considerable variation in chemical composition. In fact, the natural fats of both plant and animal origin are generally mixtures of simple and mixed triacylglycerols.

lipolysis

The general term *lipolysis* refers to the breakdown or degradation (by hydrolysis) of triacylglycerols to their constituent building blocks, glycerol and free (unesterified) fatty acids (FFA). This process occurs under the following circumstances:

1. when the body mobilizes the fat stored in adipose tissue for energy; this is *endogenous fat,* or fat from sources *within* the body, and
2. when fats taken in the diet undergo digestion in the gastrointestinal tract; this is *exogenous fat,* or fat from sources *outside* the body.

endogenous

exogenous

Lipolysis is mediated by a variety of digestive and cellular enzymes called *lipases*. The reverse process, in which cells synthesize triacylglycerols from the appropriate precursors, is termed *lipogenesis*. These two aspects of fat metabolism in the body, degradation and synthesis, can be summed up as follows:

lipases

lipogenesis

Lipolysis

$$\begin{matrix}\text{Stored or Dietary} \\ \text{Triacylglycerol}\end{matrix} + 3H_2O \xrightarrow{\text{lipase}} 3 \text{ Fatty Acids} + \text{Glycerol}$$

Lipogenesis

$$3 \text{ Fatty Acids} + \text{Glycerol} \xrightarrow[\text{intracellular enzymes}]{\text{various}} \text{Triacylglycerol} + 3H_2O$$

PHOSPHOLIPIDS AND GLYCOLIPIDS

phospholipids

Phospholipids are key structural components of biological membranes (see Chap. 9). Most membrane phospholipids are glycerol-based compound lipids or *phosphoglycerides* (Fig. 6-3). These are essentially composed of a phosphatidic acid unit, that is, a diacylglycerol with a phosphate group on the third hydroxyl group of the glycerol molecule. The phosphate group usually forms a second ester linkage with one of several alcohols, such as choline, ethanolamine, or inositol. Two widely distributed phospholipids in human cells are phosphatidyl choline, commonly known as *lecithin,* and phosphatidyl ethanolamine, commonly known as *cephalin.* A second species of membrane phospholipid, *sphingomyelin,* is derived from the long-chain (18 carbon atoms) amino alcohol, sphingosine, rather than glycerol. Sphingomyelin is likewise a major membrane component; it constitutes up to 30% of the phospholipid content of some cell membranes.

lecithin
cephalin
sphingomyelin

Another important group of cell membrane lipids is the sugar-

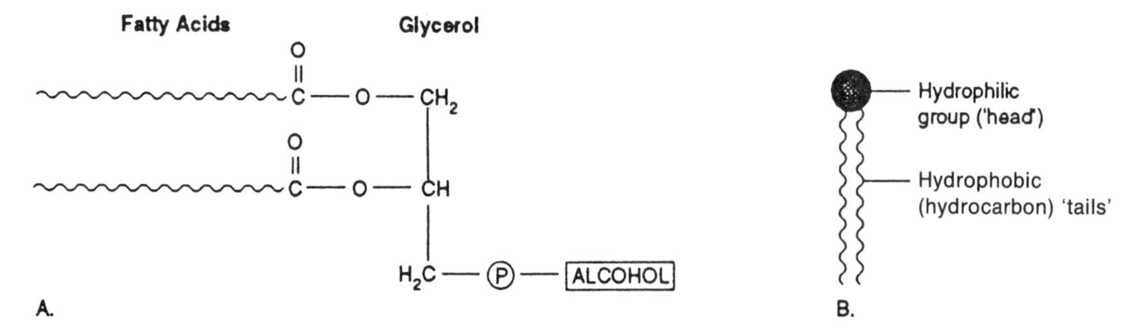

Fig. 6-3. (A) Structure of a phospholipid (phosphoglyceride). (P) symbolizes the phosphate group in the molecule. (B) Schematic representation of a phospholipid molecule. The phosphate-alcohol group is the polar hydrophilic 'head' of the molecule, and the two fatty acyl chains are the hydrophobic 'tails.'

containing *glycolipids*. These resemble sphingomyelin in that they have the sphingosine backbone, but the phosphate groups are absent. In the glycolipids, one or more sugars are directly linked to the sphingosine molecule (see also Chapt. 5). The simplest glycolipids are the *cerebrosides* which have one sugar residue, either galactose or glucose. The more complex *gangliosides* have oligosaccharide chains typically containing several amino sugars, including sialic acids.

glycolipids

cerebroside

ganglioside

Several hereditary *lipid storage disorders* of varying severity are associated with defective phospholipid or glycolipid metabolism in cells. A frequently quoted example is *Tay-Sachs disease,* a fatal condition characterized by the toxic accumulation of a type of ganglioside in neurons (nerve cells) and the degeneration of the nervous system. This is caused by the inheritance of a defective gene leading to a deficiency of the enzyme, β-hexosaminidase, which is needed for the breakdown of the ganglioside. The occurrence of Tay-Sachs disease, and of similar genetic storage diseases in man, provides important insights into what has been termed the domestic economy of the cell. Cellular synthesis of a substance, such as a phospholipid or glycolipid, is normally balanced by an equivalent rate of its breakdown or degradation. In other words, as the cell chemically destroys 'old' and 'worn-out' molecules of any given cell product, it replaces them with newly-synthesized molecules. Cellular concentrations of these materials therefore normally tend to remain constant. This dynamic balance of synthesis and degradation, of loss and renewal, defines the very important biological (and metabolic) concept of *turnover*. It may be noted here that the enzymes that degrade or digest used-up cell constituents are mainly found in cell organelles called *lysosomes,* hence *lysosomal enzymes*. The deficient enzymes in Tay-Sachs disease and other storage diseases are lysosomal enzymes. These enzymes are proteins, and their synthesis, in common with the synthesis of all types of proteins in the cell, is programmed by genes (see Chapt. 12).

lipid storage disorders

turnover

lysosomal enzymes

Amphipathic Properties

Phospholipid and glycolipid molecules have both a hydrophilic ('water-loving') and a hydrophobic ('water-fearing') region. This endows them with *amphipathic* properties (*amphi*—of both kinds). The hydrophobic units of these molecules consist of two nonpolar hydrocarbon chains or 'tails.' The hydrophilic units or polar 'heads' of the phospholipids, which have an affinity for water, are the phosphate-alcohol groups (as shown in Fig. 6-3). In glycolipids, the hydrophilic heads are the sugar residues.

amphipathic

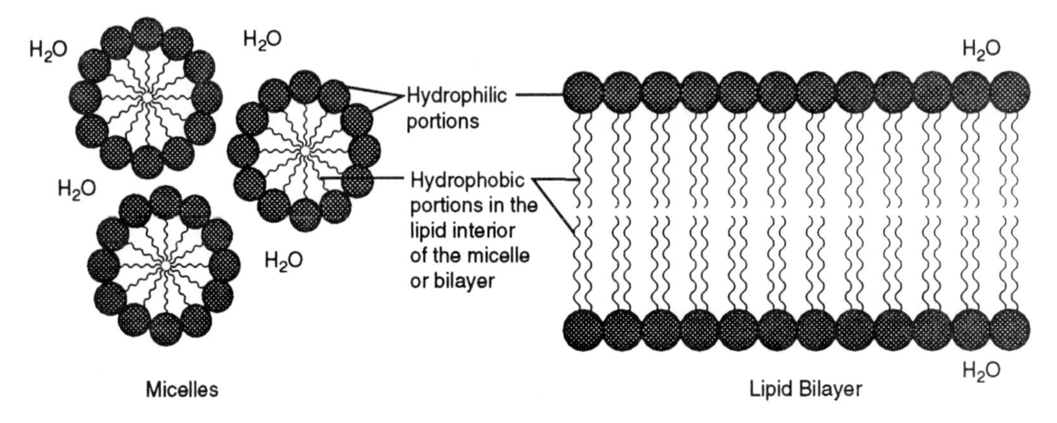

Fig. 6-4. Characteristic micelle and lipid bilayer arrangements of amphipathic molecules in aqueous media. Only a section of a lipid bilayer is shown. Intact bilayers have no free open ends; they are closed off structures.

micelles
lipid bilayer

When placed in aqueous solutions, these amphipathic molecules spontaneously arrange themselves so that their hydrophilic regions extend into the surrounding water, and their hydrophobic regions, driven by hydrophobic interactions, cluster together away from the water. Two types of arrangements occur here: the *micelle* and the *lipid bilayer* (Fig. 6-4). Both are of great significance in biological systems.

Micelles are tiny, spherical, water-soluble aggregates of amphipathic molecules, usually less than 20 nm in diameter. The hydrophilic 'heads' of the molecules face outward, exposed to the water molecules, while their fatty chains are sequestered in the interior of the micelle, avoiding contact with water. A micelle may contain hundreds or thousands of amphipathic molecules. The formation of micelles allows various water-insoluble substances to be taken up in the lipid interior of the micelle, and thus solubilized in the aqueous media of the body. This mechanism is of vital importance in the absorption of fatty compounds from the small intestine.

The lipid bilayer is the characteristic arrangement of amphipathic phospholipids and glycolipids in aqueous media. It is a sheetlike bimolecular (two molecules thick) structure. Here, once again, the fatty hydrocarbon 'tails' of the molecules are sequestered in the interior of the bilayer, shunning water, and the polar 'heads' face the aqueous solutions on both sides of the bilayer. All cellular membranes are essentially lipid bilayers formed by aggregates of phospholipids, and to a lesser extent, glycolipids. As we shall see in Chap. 9, they are self-assembling fluid structures that make highly effective barriers in the compartmentation of the cell.

Fatty acids may also exhibit amphipathic properties. In this case, the hydrophilic region is the terminal carboxyl group (—COOH) which has an affinity for water; the hydrophobic region is the hydrocarbon chain. The long hydrocarbon chains in most naturally-occurring fatty acids, however, tend to make these compounds predominantly hydrophobic and water-insoluble. On the other hand, *soaps,* which are sodium or potassium salts of long chain fatty acids (e.g., sodium stearate and sodium oleate) are amphipathic molecules that readily form micelles. *Detergents* are likewise amphipathic micelle-forming compounds. The cleansing action of soaps and detergents is due to their ability to solubilize grease or other insoluble hydrocarbon substances, and disperse them in water.

soaps

detergents

STEROIDS

The chemical structure of steroids is different from that of the lipids described above. However, since they are fatty substances found in cells and are generally insoluble in water and soluble in fat solvents, they are classed as lipids.

Steroids are complex *aromatic* compounds derived from the 3-benzene ring hydrocarbon phenanthrene (Fig. 6-5, *A*). In addition to the phenanthrene nucleus, the parent compound of the

steroids

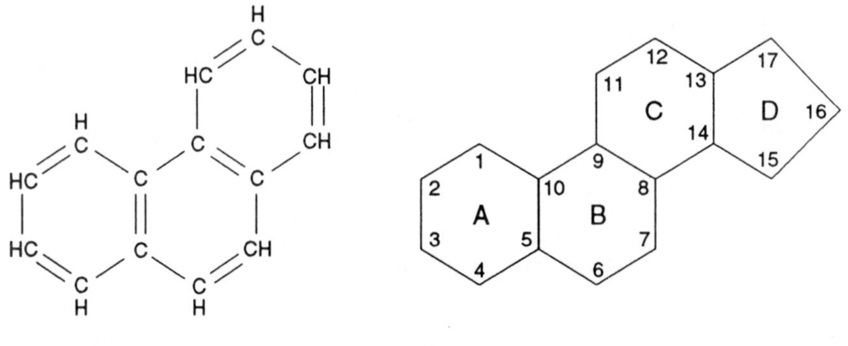

A. B.

Fig. 6-5. A, Phenanthrene, from which the steroid nucleus is derived. **B,** The perhydrocyclopentanophenanthrene nucleus, the parent compound of all biological steroids. The 5-carbon cyclopentane is indicated by the letter *D*. The position of each of the 17 carbon atoms in the steroid nucleus is shown by small numbers.

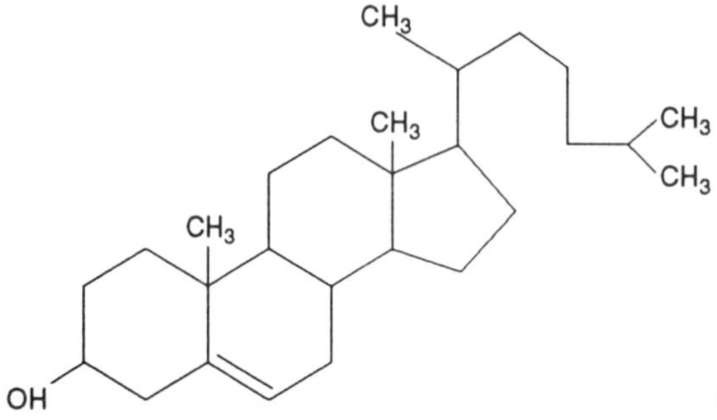

CH_3 CH_3 CH_3 CH_3 CH_3 OH

Cholesterol

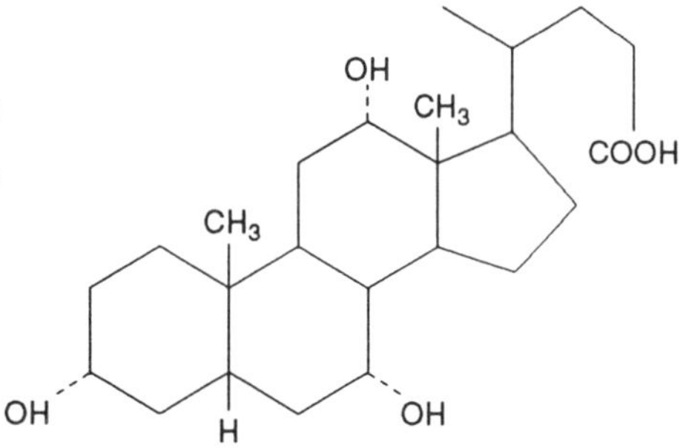

OH CH_3 COOH CH_3 OH H OH

A bile acid (cholic acid)

CH_3 CH_3 CH_3 CH_2 OH

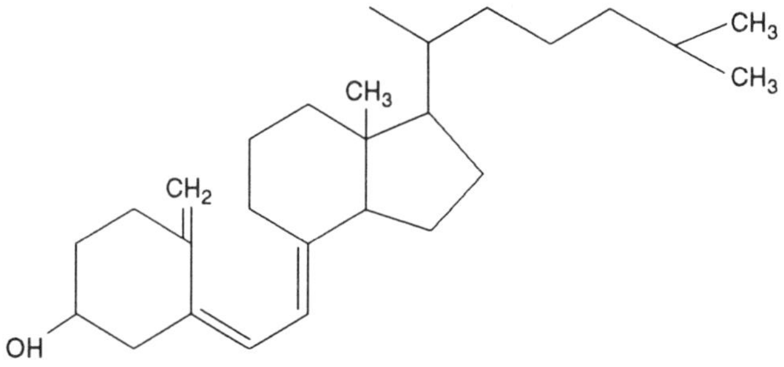

Vitamin D_3 (cholecalciferol)

The Female Sex Hormones

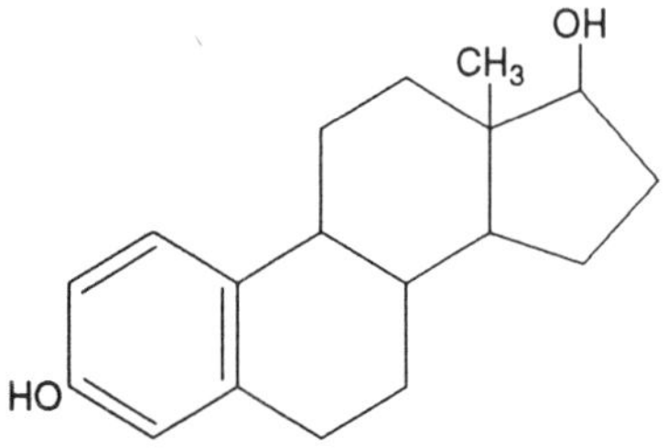

OH CH_3 HO

Estradiol (ovarian follicles)

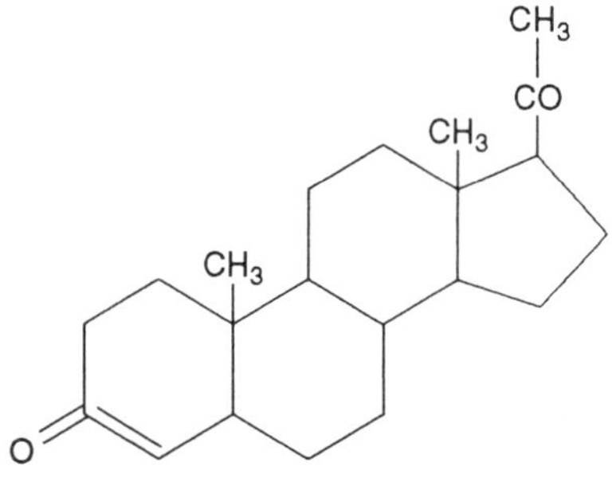

CH_3 CO CH_3 CH_3 O

Progesterone (corpus luteum)

Fig. 6-6. The steroids. (Bonds shown by dashed lines indicate that the groups project behind the plane of the page.)

The Male Sex Hormones

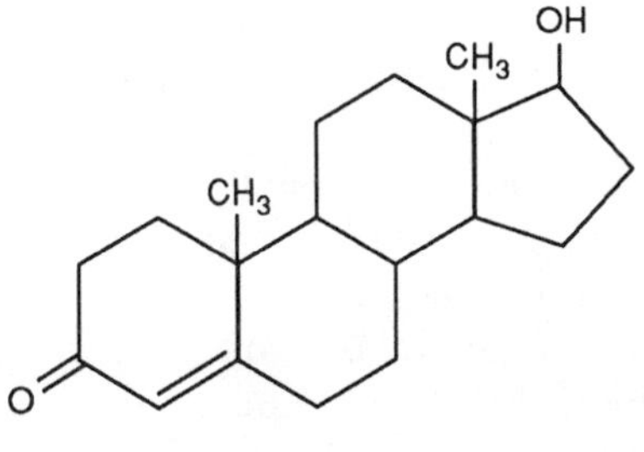

Testosterone (an androgen)

The Adrenal Corticosteroids

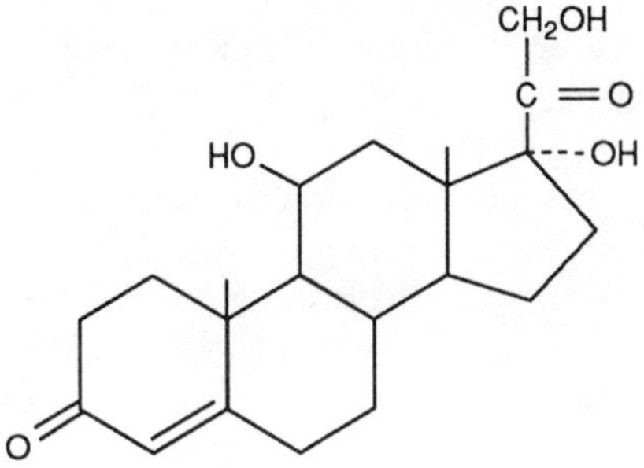

Cortisol (a glucocorticoid)

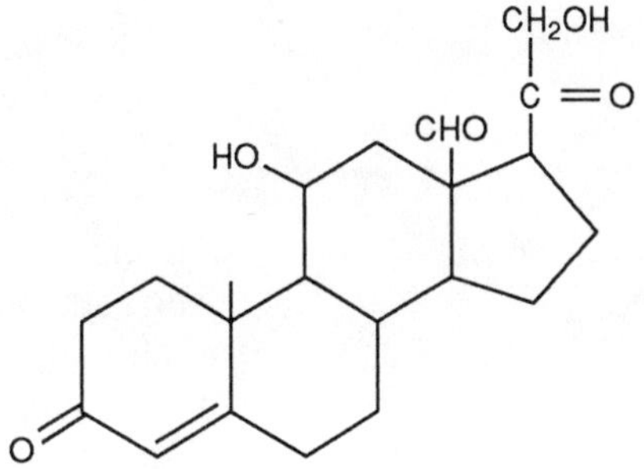

Aldosterone (a mineralocorticoid)

Fig. 6-6. *(continued)* For legend see opposite page.

steroids contains a five-membered ring (Fig. 6-5, *B*). This parent structure, from which all the biologically active steroids are derived, is called the *perhydrocyclopentanophenanthrene* nucleus. The structural formulas of the biological steroids discussed below are shown in Fig. 6-6. The illustrations show only the skeleton formula of the steroid nucleus and the important side chains.

Sterols (Steroid Alcohols)

cholesterol

The most widely distributed steroid in the body is the sterol *cholesterol*. It is found in the uncombined state, as well as in an esterfied form (combined with fatty acids), in tissues and in the lipoprotein complexes of blood plasma. Cholesterol is present in practically all dietary fats of animal origin, that is, milk fat (butter, cream, cheese), egg yolks, and the fat in meat. It is not present in plants. The compound has great physiological significance; it is an important structural component of all cellular membranes, and the cholesterol molecule is a precursor in the synthesis of the adrenocortical hormones, the male and female sex hormones, and the bile salts. The biosynthesis of cholesterol in the body, by the process of *cholesterogenesis,* is subject to feedback controls. When a sufficient amount of cholesterol is taken in the diet, cholesterol synthesis by cells (notably in the liver and the mucosa of the small intestine) is suppressed. Conversely, when the dietary intake is below the body requirements, the 27-carbon molecule is synthesized by these cells in a stepwise manner from a 2-carbon acetyl compound (the cellular metabolite, acetyl CoA; see Chapter 10). Excess dietary cholesterol has been implicated in the deposition of fatty plaques on the walls of arteries, often resulting in coronary thrombosis and stroke.

cholesterogenesis

Vitamin D is also a steroid alcohol. The precursor of vitamin D is *7-dehydrocholesterol,* present in the skin of mammals. It is converted to vitamin D_3 *(cholecalciferol)* by ultraviolet radiation (from sunlight). Cholecalciferol is also present in a few foods, notably in fish livers. Cod liver oil, for example, is a traditional dietary supplement. In the United States, the vitamin is routinely added to milk.

cholecalciferol

Cholecalciferol (vitamin D_3) is not physiologically active. The active form is synthesized in the body in the course of two sequential hydroxylation reactions. Liver cells *(hepatocytes)* carry out the first hydroxylation, adding one hydroxyl group to the molecule, and converting cholecalciferol to *25-hydroxycholecalciferol,* or more briefly, 25-OH D_3. The latter compound is then transported by way of the bloodstream to the kidney, where the second hydroxylation reaction takes place. This results in the formation of *1,25 dihydroxycholecalciferol,* or 1,25-diOH D_3, which is the most potent form of vitamin D. In the conventional sense of the term, 1,25-diOH D_3 is not a vitamin. It is classified as a hormone because it is synthesized in the body, secreted into the bloodstream, and its mode of action on target tissues is very similar to that of steroid hormones in general. The active form of vitamin D plays a major role in calcium homeostasis by facilitating the absorption of calcium from the small intestine. It also promotes calcium mobilization from bone matrix. In precisely regulating plasma calcium concentrations, it acts synergistically (working together) with a hormone from the parathyroid glands, parathormone (PTH). The two hormones have an additional relationship in that parathormone increases the rate of synthesis of 1,25-diOH D_3 in the kidney. A deficiency of vitamin D causes the demineral-

25-OH D_3

1,25-diOH D_3

ization of bone, which leads to *rickets* in growing children, and *osteomalacia* (softening of the bones) in adults.

rickets
osteomalacia

Steroid Acids (Bile Acids)

The main avenue for the breakdown and removal of cholesterol in the body is by oxidizing it to form *bile acids*. This chemistry takes place in the liver. Hepatocytes synthesize two primary bile acids from cholesterol; they are *cholic acid*, and its close chemical relative, *chenodeoxycholic acid*. The primary acids are then conjugated (coupled together) with a molecule of either glycine or taurine, which are amino acids. These conjugates are important constituents of bile, the digestive secretion of the liver. Bile is released into the duodenum (small intestine) by way of the common bile duct. The bile acids exist in bile mainly as sodium salts, and are thus conventionally termed *bile salts*. The bile salts have both a polar (hydrophilic) and a nonpolar (hydrophobic) end, and are strongly amphipathic. They are, in effect, detergent substances that are absolutely essential for both the digestion and the absorption of lipids.

bile acids

bile salts

At body temperature, dietary fats (triacylglycerols) are liquid oils which float in the aqueous medium of the digestive tract in the form of large droplets. The fat-digesting enzymes, called lipases, cannot hydrolyze these large droplets to any significant extent. The enzymatic digestion (by hydrolysis) of triacylglycerols therefore depends on the presence of the bile salts which function in this context as *emulsifying agents*. They break up and disperse the large oily aggregates into very fine particles of about 1 μm in diameter. A suspension of this type is called an *emulsion*. The formation of the emulsion greatly increases the surface area that is exposed to the action of the lipase, and thus facilitates the digestion of triacylglycerols.

emulsifying agents

The predominantly hydrophobic products of fat digestion, namely, long-chain fatty acids and monoglycerides, together with partially hydrolyzed phospholipids, cholesterol, and the fat-soluble vitamins, A, D, E and K, cannot be absorbed by the epithelium lining the small intestine unless they are solubilized. Here again, the bile salts play a pivotal role by forming mixed micelles which, it should be noted, are much smaller than emulsified particles. The assorted lipids are incorporated into the mixed bile salt micelles, and delivered via these water-soluble aggregates to the absorptive epithelial surface.

Most of the secreted bile acids are subsequently reabsorbed from the small intestine, and returned, by way of the portal vein, enterohepatic circulation

to the liver, which re-secretes them into bile. This continual recycling of bile acids between the intestine and the liver is known as the *enterohepatic circulation*. The recycling is efficient and economical. Of the 20 to 30 grams of bile acids that enter the small intestine per day, only about 10% is lost in the feces.

Steroid Hormones

estrogens

progesterone

androgens

The sex hormones are a group of related steroids that are broadly classified as estrogens and progesterone (female sex hormones) and androgens (male sex hormones). *Estrogens*, for example, estrone and estradiol, are synthesized by the follicular cells of the ovary; *progesterone*, by the cells of the corpus luteum. They differ somewhat in their physiological actions. The most representative *androgen* is testosterone, a hormone synthesized by the interstitial cells (of Leydig) of the male testis. The sex hormones regulate reproductive functions and are responsible for the development of secondary male and female sex characteristics.

corticosteroids

cortisol

aldosterone

The *cortex* of the adrenal gland is particularly active in producing a large number of steroid hormones. The *corticosteroids* can be classified in terms of biological activity as glucocorticoids, mineralocorticoids, and sex hormones. The glucocorticoids, represented by the hormone *cortisol*, regulate carbohydrate, protein, and fat metabolism and play important roles in the body's reaction to stress. The principal mineralocorticoid is *aldosterone*, which maintains the water and electrolyte balance of the body. Small quantities of estrogens and androgens are also produced by the adrenal cortex.

Plant Steroids

digitalis

A group of steroid substances found in plants are of great pharmacological importance. The representative compound of this group is *digitalis*, a steroid present in the foxglove plant. Extracts of this plant were used for centuries as a herbal remedy for congestive heart failure. Many derivatives of digitalis are now routinely prescribed for the same purpose. The digitalis steroids have potent cellular effects on cardiac muscle fibers; the overall result is a slowing and strengthening of the heartbeat and an increased efficiency of the pumping action of the heart.

FAT-SOLUBLE VITAMINS

The vitamins are a diverse group of organic compounds that must be taken in the diet because they cannot be synthesized in the body. They are essential in relatively small amounts for a

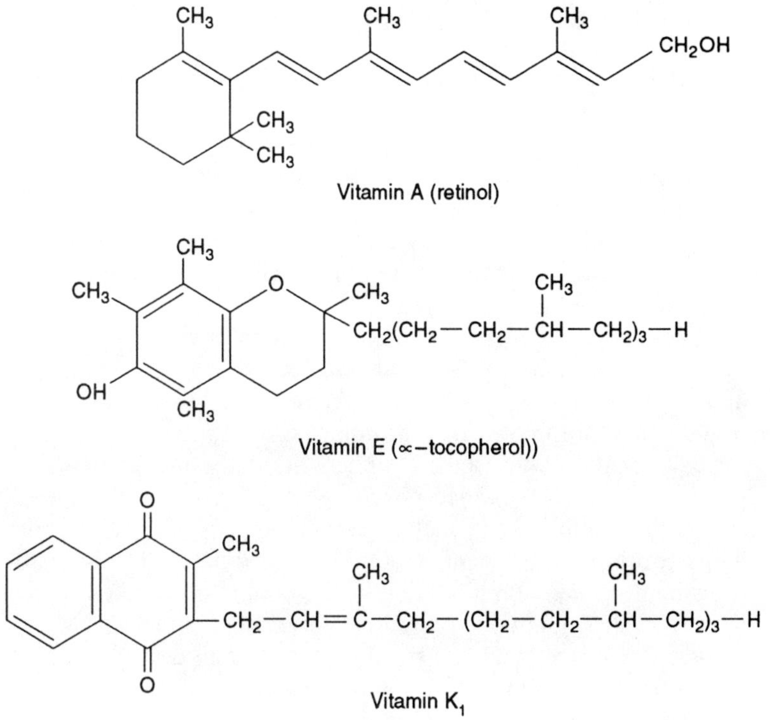

Vitamin A (retinol)

Vitamin E (∝–tocopherol))

Vitamin K_1

Fig. 6-7. Chemical structure of vitamins A, E, and K.

broad spectrum of cell functions, ranging from metabolism to differentiation. In general, vitamins are classified as either water-soluble or fat-soluble. With the possible exception of vitamin C, discussed in Chapter 5, all of the water-soluble vitamins are known to function as coenzymes (see Table 8-1, Chapter 8). The fat-soluble vitamins, A, D, E, and K, are included here in the discussion of lipid-type compounds because they are insoluble in water and can be extracted from plant and animal tissues, along with other lipids, by fat solvents. Vitamin D has a steroid structure and has been described above. Vitamins A, E, and K are essentially long-chain unsaturated hydrocarbons attached to 1 or 2 ring structures (Fig. 6-7).

retinol

Vitamin A exists in several physiologically active forms, including the primary alcohol, *retinol* (vitamin A_1), *retinal* (vitamin A aldehyde), and *retinoic acid*. The vitamin is stored in the liver and other tissues as retinyl esters; these provide a major dietary (animal) source of vitamin A. Yellow and green fruits and vegetables which contain the provitamin A plant pigment, carotene, are also an important dietary source. The most active carotenoid in plants is *β-carotene*. Ingested β-carotene is partially converted to retinol by the cells lining the small intestine. Vitamin A is required for a variety of functions in humans, i.e., a normal growth and maintenance of epithelial tissues, bone growth, reproduction and development of the embryo. Recent evidence indicates that a deficiency of vitamin A may enhance susceptibility to epithelial tumors. The best known function of vitamin A, and one that has been recognized for many years, is its role in the visual mechanisms of the retina of the eye. The light receptor cells of the retina are the *rods* and *cones*. The rods are receptors for vision in dim light or night vision (scotopic vision), and the cones for vision in bright light (photopic vision) and color vision. Both types of receptor cells contain photosensitive pigments that consist of various proteins, collectively called *opsins*, bound to *11-cis-retinal*, a derivative of vitamin A. The latter, termed the *chromophore*, is the light-absorbing component of all the photopigments. The rod photopigment, *rhodopsin*, has been the focus of many studies. When light rays strike rhodopsin, a chemical change is initiated which converts 11-cis-retinal to all-trans-retinal. The all-trans-retinal detaches from the opsin component of the molecule, causing bleaching of the pigment. These chemical changes are ultimately converted to the language of the nervous system, i.e., nerve impulses, that are transmitted to the brain, and give rise to the special sense of vision. As is obvious, synthesis of the photopigments is dependent on an adequate supply of dietary vitamin A. Indeed, one of the first symptoms of vitamin A deficiency in humans is *night blindness* (impairment of vision in dim light) which directly correlates with decreased concentrations of retinal and rhodopsin in the retina. In this case, rod function appears to be more affected by the deficiency of the vitamin than cone function.

Vitamin E activity is exhibited by several naturally occurring compounds, chemically grouped as tocopherols. The molecules exist as optical isomers, the D-forms being biologically more active than the L-forms. The most active form of vitamin E is *D-α-tocopherol*. The tocopherols were first isolated from wheat germ oil, but other vegetable oils are also good dietary sources. Although vitamin E has been shown to be essential for normal reproductive and neuromuscular function in experimental animals, its requirement in humans has been the subject of debate. In recent years, however, the *antioxidant* properties of this vitamin have attracted much attention. It is now evident that vitamin E protects cell components from the destructive oxidation effects of *free radicals*. The most common free radicals in cells are free

beta-carotene
rods and cones
opsins
chromophore
rhodopsin
night blindness
D-α-tocopherol
antioxidant
free radicals

oxygen radicals in the form of superoxide anions, hydrogen peroxide, and hydroxyl radicals. These are highly reactive chemical groups that are metabolically generated in cells under normal circumstances, but they also tend to accumulate in response to a variety of factors which include disease and injury, drugs, and exposure to environmental hazards, such as air pollution and ultraviolet radiation. The lipids of cell membranes, cellular proteins, and DNA, the genetic material of the cell, are most susceptible to the toxic impact of free radicals. Cell damage resulting from this ongoing *oxidative stress* (or wear and tear) may be implicated in a number of degenerative and age-linked diseases in humans. Antioxidants trap free radicals and inactivate them; they are, in other words, 'radical scavengers.' Vitamin E, together with vitamin C and beta-carotene, are now considered to be physiological antioxidants of major importance, and there is increasing support for the concept that they play a vital role in maintaining optimal health.

oxidative stress

Chemically, vitamin K is a derivative of *naphthoquinone*. In the absence of the vitamin, liver cells cannot synthesize the plasma protein *prothrombin* and several other coagulation factors necessary for normal blood coagulation. Thus, vitamin K deficiency can cause internal bleeding because of prolonged clotting time. Two sources of vitamin K are available to meet human requirements. The vitamin is present in a variety of plant and animal foods, and it is also synthesized by bacterial flora in the intestine. Deficiency in humans is usually caused by faulty absorption of the fat-soluble vitamin from the intestine, due to a lack of bile. Prolonged use of broad-spectrum antibiotics may also destroy the intestinal flora that synthesize the vitamin. A vitamin-K deficiency in neonates (newborn infants) sometimes occurs because of the low bacterial population of the infant intestine.

prothrombin

OTHER PHYSIOLOGICALLY ACTIVE LIPIDS: THE PROSTAGLANDINS

The *prostaglandins* were first identified in human seminal fluid in the early 1930's, and were given this name on the mistaken assumption that they came primarily from the prostate gland. The name remains even though they have been found in practically every mammalian tissue so far examined. Prostaglandins are a family of compounds, all of which are structural variants of a polyunsaturated 20-carbon fatty acid incorporating a cyclopen-

prostaglandins

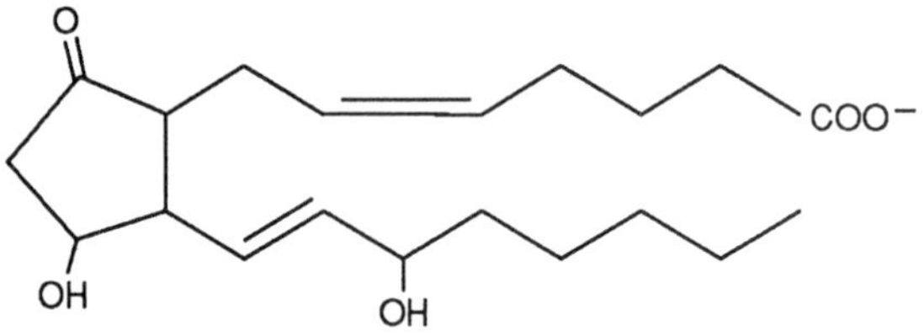

O
COO−
OH
OH

Fig. 6-8. Structural formula of a representative prostaglandin with two double bonds (PGE_2).

tane ring (Fig. 6-8). They are derived from the 20-carbon fatty acids in cell membrane phospholipids. In man, the major precursor is arachidonic acid (see Table 6-1). Members of the prostaglandin family are symbolized by the letters PG, followed by a third letter and a subscript number indicating the number of double bonds in the molecule (e.g., PGA_1, PGE_2, PGH_2, etc.). Two other groups of compounds chemically and functionally related to the prostaglandins are the *thromboxanes* and *leukotrienes*.

autocrine/paracrine

The prostaglandins are not stored in cells, but are synthesized when arachidonic acid is released from membrane phospholipids during cellular responses to various stimuli. Formerly considered to be 'local hormones,' they are more precisely classified as *autocrine/paracrine* agents, that is, locally-acting chemical signals that exert their effects on the cell that secretes them (autocrine; *auto*-self), and on cells adjacent to the secreting cell (paracrine; *para*-beside). Although they are produced in the body in minute amounts, and are rapidly degraded, the prostaglandins and related compounds are physiologically highly potent substances with far-ranging effects. Among the host of functions they mediate are immune responses, gastric secretion, uterine contraction, blood vessel tone, and platelet aggregation. In some instances, the actions of one type of prostaglandin are opposed or inhibited by those of another. For example, when a blood vessel is damaged, platelets release thromboxane A_2 (TXA_2), a prostaglandin variant that promotes the aggregation of platelets and their adhesion to the damaged site. This formation of what is termed a 'platelet plug' is an important aspect of hemostasis (prevention of blood loss) and blood coagulation. At the same time, prostacyclin (PGI_2), synthesized and released by endothelial cells lining blood vessels, acts to inhibit thromboxane and to curb excessive platelet aggregation. Thus, in the normal course of events, platelet aggregation and clot formation are controlled by a balance of opposing factors.

It has been known for some time that certain prostaglandins released by injured cells can produce inflammation (redness and swelling), pain, and fever. In the 1970's, the British pharmacologist, John Vane, demonstrated that aspirin (acetylsalicylic acid) and related compounds prevent the synthesis of prostaglandins by blocking a cellular enzyme (cyclooxygenase or prostaglandin synthase) required for the conversion of arachidonic acid to prostaglandin intermediates. This discovery finally provided an explanation of why aspirin and aspirinlike drugs (which are now generally termed 'nonsteroidal anti-inflammatory drugs,' or *NSAIDs,* for short) are so effective in the treatment of inflammatory conditions, pain and fever. Dr. Vane, who also discovered prostacyclin, was awarded the Nobel prize in 1982.

NSAIDs

SOME FEATURES OF LIPID TRANSPORT IN BLOOD

The lipids that enter the blood stream have either a dietary origin (exogenous lipids absorbed from the intestine), or they have been synthesized in tissues, notably the liver, from fatty acids and carbohydrates (endogenous lipids). Since the fluid portion of blood, that is, blood plasma, is an aqueous solution, the transport of water-insoluble hydrophobic lipids presents certain physical problems. These are solved by incorporating the lipids into *plasma lipoprotein* complexes, a group of spherical particles containing varying amounts of lipids, together with proteins called *apolipoproteins*. Each lipoprotein particle is made up of a fatty hydrophobic core of triacylglycerols and cholesterol esters which is shielded from the aqueous plasma by a surface coating of amphipathic phospholipids, apolipoprotein, and small amounts of unesterified cholesterol. The surface components have the characteristic arrangement of amphipathic molecules in water, namely, their polar groups are exposed and in contact with the surrounding aqueous medium. This neat packaging solubilizes hydrophobic lipids for transport in the blood stream. Besides contributing to the structure and solubility of these particles, the apolipoproteins interact with various enzymes and cell surface receptors, and are largely responsible for directing the metabolic processing of lipoproteins in the body.

plasma lipoproteins

apolipoproteins

Lipoprotein particles are classified according to how they behave when they are centrifuged in aqueous solutions with different specific gravities. Those that contain more fat are *lighter* (lower

Table 6-2. The plasma lipoproteins

Name	Size of particle (nm)	Components	Where formed	Comments and functions
Chylomicrons	100 or more	Mainly triacylglycerol: small amounts of cholesterol, phospholipid, and protein	Mucosal cells, small intestine	The largest lipoprotein particles; impart a milky or turbid appearance to blood plasma; main function—transport of dietary fat
Very low-density lipoproteins (VLDL)	30–80	Mainly triacylglycerol; about 10% to 15% each, of cholesterol, phospholipid, and protein	Mainly liver	Large particles; may also lend a milky or cloudy appearance to blood plasma; main function—transport of triacyclglycerol from liver to other tissues
Low-density lipoproteins (LDL)	20–30	Mainly cholesterol and cholesterol esters; about 10% to 25% each, of triacylglycerol, phospholipid, and protein	Mainly liver	Main function—transport of cholesterol
High-density lipoproteins (HDL)	5–10	Mainly protein; about 15% to 30% each, of cholesterol and phospholipid; very little triacylglycerol	Mainly liver	Main functions—transport of cholesterol to the liver for excretion in bile; source of apolipoproteins

LDL
VLDL

HDL
chylomicron

specific gravity) and are called either *low-density lipoproteins (LDL)* or *very low-density lipoproteins (VLDL),* whereas those that contain less fat and more protein are *heavier* (higher specific gravity) and are called *high-density lipoproteins (HDL)*. The largest of the plasma lipoprotein particles are the *chylomicrons*. These are assembled in cells lining the intestine, and contain about 85% triacylglycerol of dietary origin.

Table 6-2 summarizes the structure and properties of the four main classes of plasma lipoproteins, that is, chylomicrons, VLDL, LDL and HDL. It will be seen that the particles containing the highest percentage of cholesterol are the low-density lipoproteins, or LDL. Current evidence strongly indicates that high blood levels of LDL are associated with an increased risk of heart attacks (coronary heart disease). The physiological importance of cholesterol in the body has been noted in a previous section of this chapter. However, diets that are overly rich in cholesterol and saturated fats can lead to an increase in circulating LDL's, and this condition can, in turn, cause the deposition of fatty plaques on artery walls *(atherosclerosis)*. In contrast, high blood levels of high-density lipoproteins, or HDL, are associated with

atherosclerosis

a decreased risk of coronary heart disease. These lipoprotein particles remove excess cholesterol from the circulation, and transport it to the liver which clears it from the body. Considering that atherosclerotic cardiovascular disease is a major cause of death in Western society, it is not surprising that LDL's and HDL's have become major research targets in recent years.

7

PROTEINS

Living organisms contain many different types of chemical compounds. Of these, a group of giant organic molecules *(macromolecules)* known as proteins are of the utmost importance. In addition to forming the structural material of cells and tissues, they play a vital functional role as enzymes, plasma proteins, antibodies, hemoglobin, and certain hormones (such as insulin). All proteins contain the elements carbon (C), hydrogen (H), oxygen (O), and nitrogen (N). Proteins differ in this respect from carbohydrates and fats, which contain only C, H, and O. Most proteins also contain about 1% of the element sulfur (S). Certain specialized proteins contain other elements such as phosphorus (found in casein, a milk protein), iron (in hemoglobin), and iodine (in thyroglobulin, the hormone-storing protein of the thyroid gland).

macromolecules

The molecular weights of proteins range from about 10^4 to 10^7 daltons (d) or more. One can gain some idea of the size of these macromolecules by comparing their molecular masses with those of glucose (about 180 d), table salt, or sodium chloride (about 58 d), and water (about 18 d).

The building blocks of proteins are amino acids; that is, all proteins are *polymers* made up of long chains of amino acid units linked together. Each amino acid unit in a large protein molecule is called an *amino acid residue;* each unit loses something when it combines with other units, and thus is no longer a complete amino acid. The more amino acid residues, the larger the protein molecule.

polymers

amino acid residue

AMINO ACIDS: THE BUILDING BLOCKS

General Structure

The proteins of all living organisms are assembled from a basic group of 20 amino acid building blocks, listed in Table 7-1 with their three-letter abbreviations (conventionally used as symbols). An additional group of about 140 amino acids has been identified in assorted proteins. These are formed by chemical modification of one of the original set of 20 *after* it has been incorporated into the protein molecule. A few examples of the more common special amino acids found in proteins are given in Table 7-2.

Amino acids are *organic acids* that contain an additional group called the amino group. An organic acid is a compound usually

carboxyl group
amino group

possessing a *carboxyl group*, —COOH (or $-\overset{\overset{\displaystyle O}{\|}}{C}-OH$) (see Chapter 4). The *amino group* is made up of 1 nitrogen atom and 2 hydrogen atoms, and its chemical symbol is $-NH_2$ (or $-\overset{\overset{\displaystyle H}{|}}{N}-H$).

Amino acids have the following general structure:

$$\underbrace{H-\overset{\overset{\displaystyle H}{|}}{N}}_{\text{amino group}} \overset{1}{-} \underset{\underset{\displaystyle R}{|_4}}{\overset{\overset{\displaystyle H}{|_3}}{C}} \overset{2}{-} \underbrace{\overset{\overset{\displaystyle O}{\|}}{C}-OH}_{\text{carboxyl group}}$$

By convention, the carbon atom to which the carboxyl group is attached is called an *alpha* (α) *carbon*. All the amino acids found

Table 7-1. The 20 amino acid building blocks of proteins

Amino acid	Abbreviation	Amino acid	Abbreviation
Alanine	Ala	Isoleucine	Ile
Arginine	Arg	Leucine	Leu
Asparagine	Asn	Lysine	Lys
Aspartic acid (aspartate)	Asp	Methionine	Met
		Phenylalanine	Phe
Cysteine	Cys	Proline	Pro
Glutamine	Gln	Serine	Ser
Glutamic acid (glutamate)	Glu	Threonine	Thr
		Tryptophan	Trp
Glycine	Gly	Tyrosine	Tyr
Histidine	His	Valine	Val

Table 7-2. Some modified amino acids in proteins

Amino acid	Description
Hydroxylysine and hydroxyproline	Hydroxylated derivatives of lysine (Lys) and proline (Pro), found in collagen, a connective tissue protein
Cystine	Formed by a *disulfide bond* between two cysteine (Cys) residues (see Fig. 7-7); common in many proteins
Iodotyrosines and iodothyronines	Formed by various combinations of tyrosine (Tyr) with iodine (I); found in thyroglobulin, the hormone-storing protein of the thyroid gland

in proteins have their amino group attached to this alpha carbon and are therefore known as *alpha amino acids*. As shown, the alpha carbon atom forms four bonds with:

α-amino acid

1 the N of the amino group
2 the C of the carboxyl group
3 a hydrogen atom
4 a side chain of variable structure, generally designated in chemical shorthand as R

When dissolved in water at neutral or physiological pH levels (7–7.4), amino acids are not undissociated molecules. They exist as dipolar ions, or *zwitterions,* which have a positively charged pole ($—NH_3^+$) and a negatively charged pole ($—COO^-$), as follows:

zwitterions

$$^{+}H_3N—\overset{\displaystyle H}{\underset{\displaystyle R}{\overset{|}{\underset{|}{C}}}}—COO^-$$

The pH at which the molecule exists in this electrically neutral form, that is, with no *net* electrical charge, is called its *isoelectric point.* At pH levels above and below this point, the zwitterion can act respectively as either a weak acid (proton donor), or a weak base (proton acceptor).

isoelectric point

Side Chains

Each of the 20 amino acids has a specific side chain that differentiates it chemically from the other amino acids. In the simplest amino acid, glycine, the R side chain stands for 1 hydrogen atom:

```
      H  H  O
      |  |  ||
   H—N—C—C—OH
         |
         H↖(R)
```

The rest of the amino acids have side chains consisting of organic groups of varying size and complexity (Fig. 7-1). The side chains of alanine, valine, and leucine, for example, have nonpolar hydrophobic side chains, while those of other amino acids, such as aspartic acid and lysine, have polar or charged side chains. Phenylalanine, tyrosine, and tryptophan have aromatic side chains (benzene ring derivatives). Methionine and cysteine have sulfur-containing side chains. As we shall see, the interactions of the various amino acid side chains in a protein molecule are largely responsible for its unique three-dimensional shape or *conformation*.

conformation

Optical Activity

Glycine excepted, all the amino acids have an asymmetrical carbon atom, with four different groups attached to it. They are thus optically active compounds; that is, they are able to rotate a plane of polarized light to the right or to the left. Like the monosaccharides (see Chapter 5), optically active amino acids have mirror-image D and L enantiomers. However, *the proteins of all living organisms contain only L-amino acids*. The few D-amino acids that do occur naturally are not incorporated in proteins; for instance, the cell walls of bacteria are made up of complex polysaccharides that contain D-alanine and D-glutamine residues. Several antibiotics, also produced by microorganisms, contain D-amino acids; for example, D-phenylalanine is found in the antibiotic gramicidin.

L-amino acids

Essential and Nonessential Amino Acids

Some amino acids, the *nonessential amino acids,* can be synthesized in the body from other compounds, and are therefore not required in the diet. However, at least nine of the basic group of twenty amino acids cannot be synthesized by humans, and must be supplied in the diet; the latter are termed *essential amino*

nonessential amino acids

essential amino acids

Group	Amino acid	Side chain (as printed, top to bottom)
Hydrogen atom	*Gly*	H
Hydrocarbon chains	*Ala*	CH_3
	Val	CH; H_3C, CH_3
	Leu	CH_2; CH; H_3C, CH_3
	Ile	$HC-CH_3$; CH_2; CH_3
Containing hydroxyl (OH) groups	*Ser*	H_2C-OH
	Thr	$HC-OH$; CH_3
Containing additional charged carboxyl group (acidic amino acids)	*Asp*	CH_2; COO^-
	Glu	CH_2; CH_2; COO^-
Amide ($O=C-NH_2$) derivatives of acidic amino acids	*Asn*	CH_2; $O=C-NH_2$
	Gln	CH_2; CH_2; $O=C-NH_2$
Containing additional charged amino groups (basic amino acids)	*Lys*	CH_2; CH_2; CH_2; CH_2; NH_3^+
	Arg	CH_2; CH_2; CH_2; NH; $C=NH$; NH_3^+
	*His**	CH_2; $C=CH$; ^+H_2N, N; C H
Containing a sulfur (S) atom	*Cys*	CH_2; SH
	Met	CH_2; CH_2; S; CH_3
Aromatic (ring) side chains	*Phe*	CH_2; benzene ring
	Tyr	CH_2; benzene ring; OH
	Trp	CH_2; C; CH; N H (indole ring)
Imino acid† (complete molecule shown with side chain bonds)	*Pro*	H; $H_2N^+-C-COO^-$; H_2C, CH_2; C H_2

Fig. 7-1. Amino acid side chains (R groups). * The heterocyclic ring in the histidine side chain is called an *imidazole* ring (see Chapter 4). The ring is usually positively charged and acts as a base (proton acceptor). † The R group of proline is bonded to both the amino group *and* the α-carbon atom. Thus, strictly speaking, proline and its derivatives are *imino,* rather than amino, acids.

acids (Table 7-3). Note that arginine can be synthesized in adequate amounts by adults, but it is usually classed as essential for growing children. Two nonessential amino acids, tyrosine and cysteine, are synthesized from the essential amino acids, phenylalanine and methionine, respectively. If inadequate amounts of

Table 7-3. Essential amino acids

Arginine*	Methionine
Histidine	Phenylalanine
Isoleucine	Threonine
Leucine	Tryptophan
Lysine	Valine

* Arginine is considered to be essential for growing children.

these precursor amino acids are taken in the diet, tyrosine and cysteine can become, in effect essential amino acids.

Dietary proteins with a high biological value are those that contain a balanced content of essential amino acids. Most proteins of animal origin (meat, fish, milk, eggs, and cheese) are in this category. In contrast, plant proteins, such as zein in maize and gluten in wheat, are lacking in one or more essential amino acids. Mixing proteins from different sources in a varied diet ensures an adequate intake of essential amino acids. Diseases due to deficiencies of essential amino acids generally appear in populations that subsist mainly on carbohydrates.

Functions of Amino Acids

In addition to being utilized by the body as building blocks of peptides and proteins, amino acids also function as precursors of neurotransmitter substances,* hormones, pigments, vitamins, and a host of miscellaneous nonprotein biomolecules. Several physiologically active amino acids in this category are never incorporated into proteins. These include the amino acids ornithine and citrulline, which are important reactants in the formation of urea by the liver, and some amino acids in which the amino groups are found on carbon atoms other than the alpha (α) carbon. Two examples of the latter are beta (β) alanine, present in the vitamin pantothenic acid, and the neurotransmitter gamma (γ) aminobu-

* Neurotransmitter substances are synthesized by neurons (nerve cells). They are the chemical messengers by which neurons communicate with each other and with the muscles and glands they innervate. A nerve impulse (an electrical wave) is the signal for the release of these substances from the terminal branches of neurons. Many of the currently recognized neurotransmitters are either amino acids or are chemically derived from amino acids.

tyric acid (GABA), which is secreted by certain neurons in the central nervous system.

FORMATION OF PEPTIDES AND POLYPEPTIDES

Peptide Bonds

peptide bonds

Amino acids link together (polymerize) by forming *peptide bonds*. A peptide bond is a covalent linkage between the carboxyl group (COOH) of one amino acid and the amino group (NH_2) of another. Fig. 7-2 indicates how *two* amino acids combine. As the peptide bond is formed, a water molecule is removed from the α-carboxyl group of one amino acid and the α-amino group of the other. The remnant of the carboxyl group is attached to the remnant of the amino group by the peptide bond:

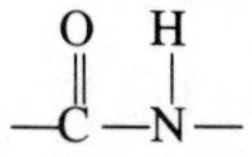

$$-\overset{\overset{\displaystyle O}{\|}}{C}-\overset{\overset{\displaystyle H}{|}}{N}-$$

Recall from Chapter 4 that just as the peptide bond is formed by the removal of a molecule of H_2O, it can likewise be broken by the addition of a molecule of H_2O, namely, by *hydrolysis*. In this

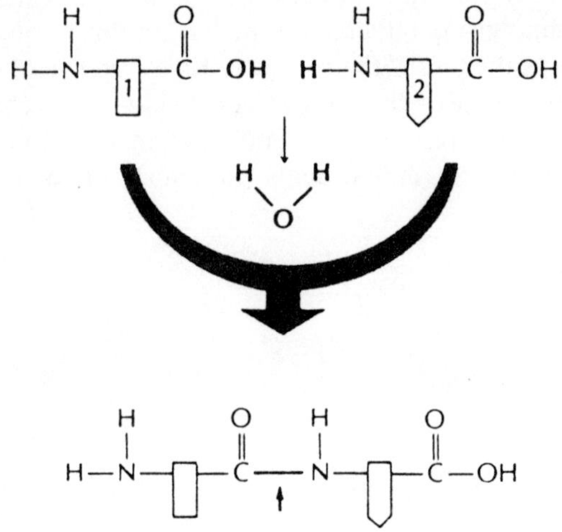

Fig. 7-2. Formation of a peptide bond (arrow) between two amino acids. (This version is simplified for clarity. The amino acids actually combine as zwitterions.)

process, each linked amino acid is released intact as it recovers its amino H and carboxyl OH portions. This is, in fact, how peptides and polypeptides (proteins) are digested or broken down in the body into their component amino acid units.

peptide
polypeptide

The terms, *peptide* and *polypeptide,* are both used to denote chains of amino acids linked together by peptide bonds. Peptides are the smaller members of the family. They are relatively short linear chain compounds, usually containing less than 50 amino acid residues. Conventionally, the number of amino acids in the peptide is designated by a numerical prefix. For example, the peptide in Fig. 7-2 is a dipeptide (containing 2 amino acid residues), while a tripeptide would have three residues, a tetrapeptide, four residues, a pentapeptide, five residues, and so on. The polypeptides, which will be considered below, are longer chains consisting of from 50 to 500 or more amino acid residues (*poly*-many). They differ from peptides in their size as well as their conformational properties.

The structural formula in Fig. 7-3 shows a tetrapeptide, composed of four amino acids (glycine, alanine, serine, and valine) linked together by peptide bonds. This tetrapeptide would be called glycylalanylserylvaline. The dashed lines bisect the peptide bonds linking the four residues together. At the formation of each of the peptide bonds, 1 molecule of H_2O is released. The characteristic side chains of each of the four amino acids are indicated by R_1, R_2, R_3, and R_4. By convention, a peptide chain, no matter how long it may be, is considered to have two different ends. The *beginning* of the chain is always the amino acid residue with an intact α-amino group (glycine on the left, in this particular example), whereas the *end* is the residue with a free α-carboxyl group (as shown by valine on the right). Each residue in the chain, except the very last one, has its name ending changed to -yl.

Like some of the amino acids previously cited, peptides are

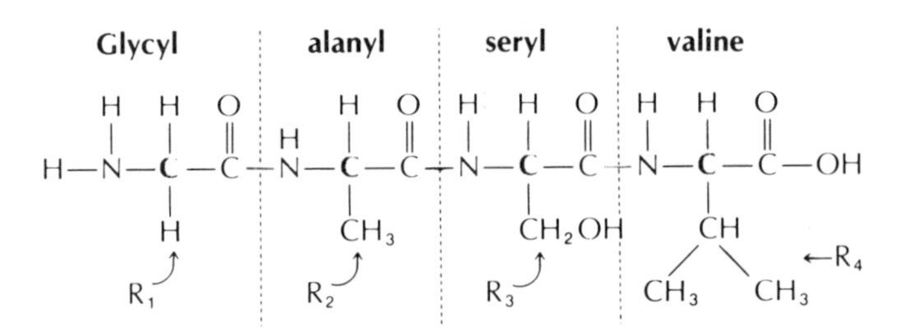

Fig. 7-3. A tetrapeptide called glycylalanylserylvaline. Peptide bonds are bisected by dashed lines. Characteristic side chains of amino acid residues are indicated by R_1, R_2, and R_4.

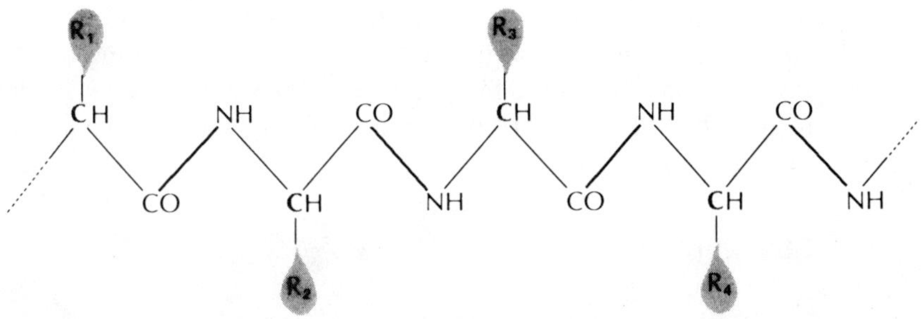

Fig. 7-4. Two-dimensional diagrammatic representation of a polypeptide chain. Alpha carbon atoms and peptide bonds are shown in bold.

per se physiologically active compounds. They have important roles in the body that are not directly associated with the formation of proteins. For example, numerous hormones are peptides, such as the *releasing* and *inhibiting hormones* of the hypothalamus, and the *gastrointestinal hormones* secreted by cells of the stomach and small intestine. Additionally, diverse peptide neurotransmitters, or *neuropeptides,* have been identified in the nervous system in recent years; they include *substance P* and the *endogenous opioids.*

releasing/inhibiting hormones

neuropeptides

Polypeptides, as previously noted, are longer chain compounds that generally contain 100 or more amino acids. They form the basic fabric of protein molecules. A polypeptide chain is a linear unbranched array of many linked amino acid units. As depicted in Fig. 7-4, the structural backbone of the chain consists of a regularly repeating sequence of alpha-carbon atoms and peptide-bonded carboxyl-amino (CO—NH) groups. The variable parts of the molecule are the side chains, R_1, R_2, etc., jutting out from the backbone.

PROTEINS

A protein molecule is made up of one or more polypeptide chains. In proteins consisting of more than one polypeptide chain, each chain is called a *subunit.* Based on the number of subunits they contain, multisubunit proteins are often described as dimers (2 subunits), trimers (3), tetramers (4), etc. Subunits in a protein may be identical polypeptide chains, or they may be different. For example, the most common type of hemoglobin molecule in adults (HbA) is a tetramer with two subunits of one kind, the alpha (α) chains, and two of another, the beta (β) chains.

subunit

Living cells synthesize all the remarkable varieties of proteins they require from the basic set of 20 amino acids listed previously. Although this may seem like a small quantity of building blocks to start with, an astronomical number of possible proteins can be formed by the assembly of the 20 different units on chains of varying length.

Structure

The chemical and biological properties of a given protein depend on the structure of its molecule. Protein structure is conventionally classified in terms of four levels or hierarchies of organization known as primary, secondary, tertiary and quaternary structures. Each succeeding level denotes an increasing order of complexity.

primary structure
amino acid sequence

Primary structure. The specific *sequence of amino acid residues* in a protein molecule determines its primary structure. This sequence of residues, that is, the position of each amino acid relative to others in the chain is essentially what makes one protein different from another. Thus, a hypothetical protein with a chain containing the sequence

Gly-Asp-Lys-Arg-Gly-Leu

will be entirely different chemically from another such protein with a chain sequence

Gly-Tyr-Lys-Arg-Gly-Leu

Note that only *one* amino acid has been changed in the sequence. The differences in chemical properties result from the differences between the respective side chains of *Asp* and *Tyr*. Given the standard group of 20 amino acids, it has been estimated that there could be about 10^{195} possible sequences for a protein containing 150 amino acid residues! However, every molecule of any given species of protein will always have the identical sequence of residues, and hence identical physical, chemical and biological properties.

The sequence of amino acids in a protein molecule may be regarded as the map of that protein. Every protein has a map unique to itself that differentiates it from others. As we shall see later on, the maps are predetermined, transmitted from one generation to the next by means of a *genetic code* that embodies all the information necessary for assembling the thousands of different protein molecules in the body, amino acid by amino acid.

genetic code

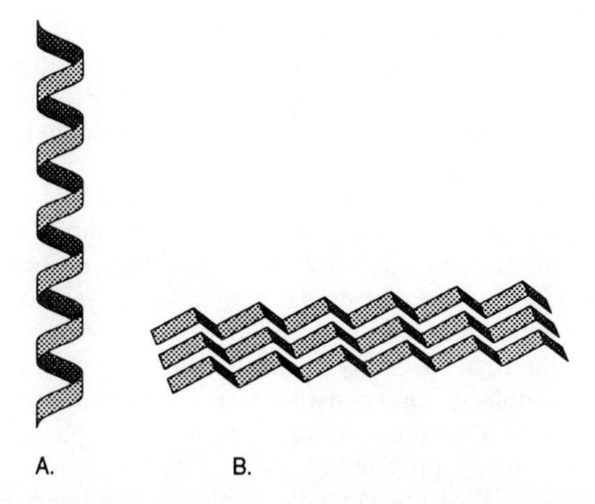

Fig. 7-5. Two common types of secondary structure found in protein molecules. **A.** The alpha helix. **B.** The beta pleated sheet.

secondary structure

Secondary structure. The polypeptide chains of proteins are flexible, and they contain many chemically reactive groups. Therefore, they do not exist simply as one-dimensional linear (primary) strands. Local interactions between nearby amino acids cause the chain to fold up, giving rise to certain spatial arrangements known as secondary structure. Two examples of secondary structure that are commonly found in proteins are the *α-helix* and the *β-pleated sheet* (Fig. 7-5, A and B). These structures are stabilized by noncovalent hydrogen bonds. The bonds form whenever a hydrogen atom of an amino group happens to lie fairly close to the oxygen of a carboxyl group (Fig. 7-6). Recall that although these bonds are individually weak, they gain strength from their numbers. Shorter or longer segments of helices and pleated sheets are found in most protein molecules, usually alternating with less regular structural arrangements. However, in a few instances, the

α-helix
β-pleated sheet

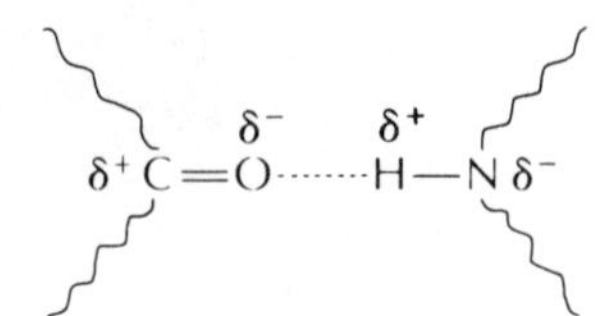

Fig. 7-6. Formation of a hydrogen bond between a carbonyl oxygen atom and an amino hydrogen. This weak attraction exists because oxygen and nitrogen atoms are stongly *electronegative*. (See Chapter 2 for discussion of hydrogen bonding.)

α-helix or β-sheet is the predominant structure in the protein. For example, the single polypeptide strand of myoglobin, an iron-containing protein found in muscle tissue, is arranged almost entirely in α-helical segments, while the β-sheet formation predominates in fibroin, the protein of silk.

tertiary structure
conformation

Tertiary structure. The complete three-dimensional shape or *conformation* of a polypeptide chain in a protein molecule is referred to as its tertiary structure. The secondary elements in a protein tend to fold up and organize themselves into compact, three-dimensional units or supersecondary structures, called *domains*. It is the final assembly or arrangement of these domains in the protein molecule that constitutes its tertiary structure. Specific domains are often the working parts of the protein molecule, and are thus directly involved in the type of function any given protein performs. With respect to the evolution and diversification of proteins over a span of some 3 billion years, it is of interest that many quite unrelated proteins have the same or very similar domains. This raises the possibility that different classes of proteins may have evolved from a common precursor.

domains

globular proteins
fibrous proteins

On the basis of their overall three-dimensional shape, proteins may be classified as either *globular proteins* or *fibrous proteins*. Globular proteins are highly coiled molecules with a roughly spherical shape (Fig. 7-7A). The plasma proteins, hemoglobin, and most enzymes are globular proteins. The molecules of fibrous proteins resemble fibers, as their name implies, that is, they have rodlike shapes. In general, these are the structural proteins of the body, such as collagen and elastin (the proteins of connective

Fig. 7-7. Examples of tertiary and quaternary structures. **A.** Tertiary structure of a globular protein. **B.** Quaternary structure of the hemoglobin molecule, showing the three-dimensional 'package' of four polypeptide chains and four iron-containing heme groups.

tissues, i.e., tendons, ligaments, and bones), and keratin (the sulfur-containing protein of skin, hair, and nails). In some proteins, notably *myosin* (a contractile protein present in the cells of muscle tissue and in other cells in the body), the molecule features both globular and fibrous regions.

quaternary structure

Quaternary structure. Multisubunit proteins that are composed of more than one polypeptide chain have a more complex level of organization called quaternary structure. The subunits are never randomly grouped, but are fitted together in a compact symmetrical 'package.' The tetrameric hemoglobin molecule was the first protein to have its quaternary structure defined (Fig. 7-7B). Quaternary structure enhances the functional effectiveness of many proteins.

Chemical Forces that Determine Protein Structure

Most of the chemical interactions that mediate and stabilize the characteristic folds and coils of polypeptide chains are noncovalent. These include hydrogen bonds, electrostatic (ionic) attractions between negatively charged and positively-charged side chains on amino acids (such as lysine, aspartic acid, etc.; refer also to Fig. 7-1), and hydrophobic interactions. With respect to the latter, recall from Chapter 3 that nonpolar hydrophobic ('water-fearing') groups in molecules tend to cluster together in an aqueous environment, shunning contact with water. On the other hand, hydrophilic ('water-loving') groups readily interact with water dipoles. Amino acids vary in their affinity for water mainly because of the properties of their side chains. For example, amino acids, such as aspartic acid, serine, lysine, and arginine, have hydrophilic charged or polar side chains with an affinity for water. In contrast, the side chains of other amino acids, such as valine, leucine, and phenylalanine, are nonpolar and hydrophobic. Since most proteins are immersed in the aqueous fluids of the body, the presence of specific sequences of amino acids in the protein molecule will cause it to fold up three-dimensionally in such a way as to segregate those residues with hydrophobic side chains in the interior of the molecule, while residues with polar side chains are exposed on the surface.

Taken individually, these noncovalent forces are relatively weak, but their combined effects become significant when large numbers of interacting groups are packed together at close range in a protein molecule.

disulfide bond

There is one *covalent* side chain interaction that frequently occurs in proteins. This is the *disulfide bond,* a strong form of

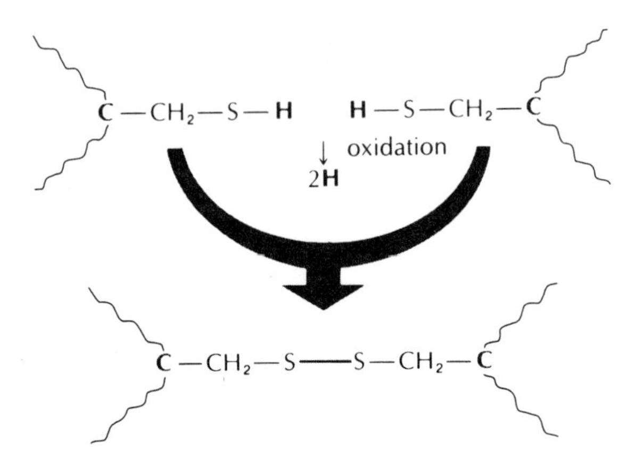

Fig. 7-8. Formation of disulfide bond by simultaneous oxidation of spatially close—SH groups of different cysteine residues. The modified amino acid that is formed by this bond is called cystine.

bonding that can cross-link different regions of a protein molecule. Note that the amino acid, cysteine, has a sulfhydryl group (SH) on its side chain. If two cysteine residues on a polypeptide chain come into fairly close proximity to each other, they will combine to form the disulfide (S-S) linkage (Fig. 7-8). The bonding produces the modified amino acid, cystine.

Cofactors and Prosthetic Groups

cofactor
prosthetic group
apoprotein

Functional proteins, particularly those that are enzymes, often contain nonpeptide components that enhance their activity. These are usually called *cofactors* if they are loosely bound to the protein part of the molecule, and *prosthetic groups* if they are tightly bound. However, the distinction between the two terms is not rigorous. A protein molecule without its prosthetic group is called an *apoprotein* (or, if it is an enzyme, an *apoenzyme*). Common cofactors in many enzymes are metal ions (Chapt. 8). The classical example of a prosthetic group is the iron-containing porphyrin ring (the *heme* group) that is present in hemoglobin, myoglobin, and a family of mitochondrial enzymes called cytochromes.

Acid/Base Properties of Proteins

Like the amino acids, proteins can act either as weak acids or weak bases, due to the presence of intact amino groups (NH_2) and carboxyl groups (COOH) at the two ends of the polypeptide chains, and various additional acidic and basic groups on the side chains of the amino acid residues. As noted previously, proteins exist mainly as weak acids at the prevailing pH of body fluids and thus make effective *buffers* (see discussion of buffers in Chapter 3). Since body fluids, such as intracellular fluid and plasma, contain significant amounts of proteins, their buffering capacity is responsible to a large extent for the maintenance of a constant physiological pH in the internal environment.

protein buffers

Also in common with their structural components, the amino acids, protein molecules are electrically neutral at the pH that constitutes their *isoelectric point*. At pH levels above and below this point, which is specific for each protein, and depending mainly on the acidic and basic R groups of the particular amino acid residues present in the protein, the protein molecules become either negatively or positively charged. Like any other charged ions, these macromolecular anions and cations will migrate in an electrical field toward the anode or the cathode. This property of proteins is the basis for the technique known as *electrophoresis*, which is employed to separate individual proteins in a mixture by taking advantage of the different electrical charges they acquire at any given pH.

isoelectric point

electrophoresis

Denaturation

The three-dimensional coiling in protein molecules can be disarranged or broken without changing the primary structure of the molecule. Any treatment, chemical or physical, that *alters the conformation* of the molecule, but does not chemically break the peptide bonds in the molecule, is said to have *denatured* the protein. Denaturation is a process by which a protein molecule is transformed from its usual orderly arrangement to an amorphous (shapeless) form (Fig. 7-9). Perhaps the best example of an irrever-

denaturation

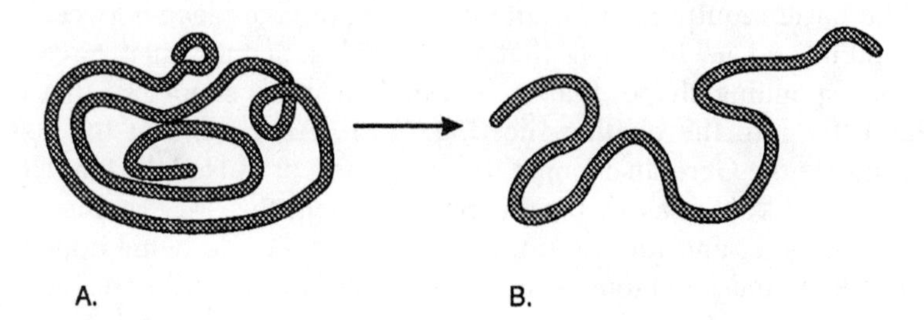

Fig. 7-9. Denaturation of a protein. **A.** The native functionally-active (coiled) form of the molecule. **B.** The denatured (unfolded) inactive form of the molecule.

sible denaturation of protein is the conversion of fluid egg albumin to a solid mass by boiling. This type of denaturation is called *coagulation*. Most proteins coagulate at high temperatures.

native state

When proteins are denatured, they lose their unique characteristics and biological activity, i.e., what is termed their *native state* is disrupted. Denaturation can be brought about by variations in temperature and pH or by treatment of the protein with organic solvents such as alcohol, precipitation of the protein with heavy metals, or exposure to ultraviolet light. In some circumstances, denaturation is partial and can be reversed.

Binding Proteins

Proteins subserve a remarkable range of structural and functional roles in living organisms (Table 7-4). For the remainder of this chapter, however, the topic of discussion will be the proteins that make their contribution to life processes by selectively binding other molecules. The activities of this large and diverse class of proteins are the molecular basis of physiology (the science of function).

binding site

ligand

Among other things, binding proteins function as enzymes, antibodies, receptors, membrane carriers, and transport proteins. Although all the molecules of any given species of binding protein have the same conformation, it is the great diversity of conformations exhibited by different binding proteins that enables them to have such a wide variety of functions. Indeed, these proteins fully exploit all possible permutations of amino acid sequences to achieve their conformational diversity. Each type of protein molecule has one or more uniquely-shaped regions on it, generally termed its *binding site(s)*. The site will 'recognize' and attach another molecule (or part of a molecule); the bound molecule is generally termed the *ligand* of that binding protein (*ligare*-bind). The basic requirement for an interaction to take place between a protein and its ligand is that the ligand molecule must have a corresponding shape, that is, a complementary shape, so that it can 'fit' into the binding site (Fig. 7-10). At the end of the last century, the German chemist, Emil Fischer, proposed the familiar 'lock and key' analogy for this phenomenon, the 'lock' being the binding site, and the 'key' the particular molecule being bound. The lock-and-key model was originally used to describe the binding activities of enzymes, and we shall consider the special characteristics of this very important category of binding proteins in detail in the next chapter.

Another model that is relevant to the interactions between pro-

Table 7-4. Biological roles of proteins

Type of function	Examples
Mechanical (structural) support, protection	Keratin (Skin & Skin Appendages) Collagen (Tendons, Ligaments, Bone) Elastin (Ligaments) Fibrin (Substance of blood clots) Histones (Structural scaffolding around DNA)
Enzymes	Many varieties
Hormones	Hormones of the anterior pituitary, kidney, placenta, etc.
Storage:	
Nutrient storage (for developing young animals)	Casein (in milk) Egg albumin
Hormone storage	Thyroglobulin
Oxygen storage	Myoglobin
Immunity	Antibodies Interferons Interleukins Complement system Histocompatibility antigens (e.g. blood group antigens)
Transport proteins:	
In Blood	Plasma albumins, globulins Hemoglobin (O_2/CO_2 transport)
On Cellular Membranes	Carriers Ion channels Ion pumps
Cellular Receptors	Receptors for hormones, neurotransmitters, and other chemical messengers
Calcium-binding Regulatory Proteins	Calmodulin, Troponin C
Cytoskeletal and Contractile Proteins (cell mobility, muscle contraction)	Actin Myosin Tubulin
Blood Coagulation Factors	Prothrombin Fibrinogen, etc.
Gene Control	Repressor/Activator proteins
Growth Factors (regulating mitosis, cell growth, cell differentiation)	Epidermal Growth Factor Nerve Growth Factor Platelet-Derived Growth Factor

teins and the molecules they bind is the surface relationship between two matching pieces of a jigsaw puzzle. These pieces fit together only because they have matching surface contours, that is, if there is a depression on one piece, there must be a protuberance on the other that is shaped so as to fit into the depression. Three-dimensionally speaking, the two pieces must be stereospe-

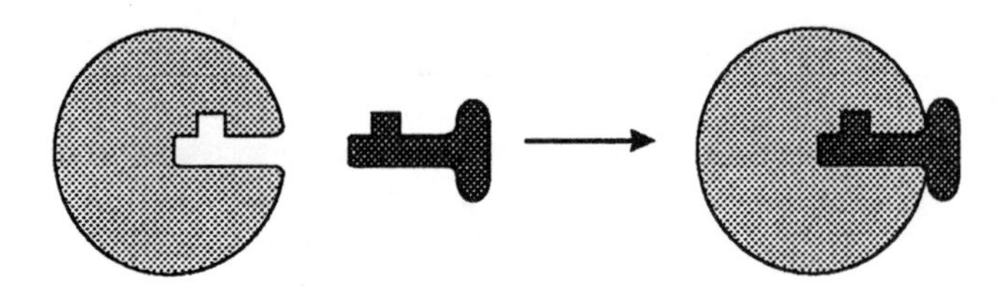

Fig. 7-10. The interaction of a binding protein with another molecule (schematically shaped like a key). The binding site of the protein (the 'lock') and ligand molecule (the 'key') have complementary shapes.

cific. It can now be understood more clearly why the right-handed optical isomer of an optically active molecule might have difficulty interacting with a protein, for example, an enzyme whose binding site happened to be complementary only to left-handed forms of that molecule. Such a situation would be comparable to trying to fit the right hand into a left-handed glove. However, it should be borne in mind that protein molecules do not really resemble rigid jigsaw puzzle pieces; they are flexible and dynamic structures.

The interactions between proteins and the molecules they bind involve noncovalent bonds that form rapidly and spontaneously. The bonds are transient and readily reversible, so that molecules are continually being bound and released from the protein at rates that depend on their concentrations, and on the availability of binding sites at any given instant.

Binding proteins may be considered signal-recognizing units in a chemical communications system where the signals are the shapes of other molecules. Since these proteins are highly specific for the molecules they bind, recognizing only one, or at the most, a very limited number of similarly shaped molecules, this type of communications system operates as a division of labor on an exceedingly fine scale. Obviously, even the most primitive unicellular organism will need a large number of different functional proteins in order to carry on all of the processes essential to its survival. For example, the common bacterium, *Escherichia coli,* has some 3000 different species of proteins.

Protein-ligand interactions mediate a very diverse range of physiological mechanisms at a molecular level. It is relevant at this point to describe briefly a few representative examples. Some proteins do little more than bind certain molecules for transport or other purposes. For example, hemoglobin binds and transports oxygen from the lungs to tissues in all parts of the body. Similarly, carrier proteins on cell membranes transport various substances into and out of cells. Antibodies, which belong to a class of proteins called immunoglobulins, bind 'foreign' molecules (antigens), and by doing so, inactivate or neutralize them. In many instances, however, the binding of a molecule to a protein acts as a trigger or a switch that sets one or another physiological process into motion. The bound molecules in these interactions act as signals; the proteins that bind them translate the signals into biological responses. The initial event here is the conformational change induced in the protein molecule when it binds another molecule. Such shape-changing events constitute *meaningful information* in the molecular language of the cell. Enzymes and cellular receptors are important examples of the latter category of binding proteins, and their activities have far-reaching effects on cell function. In the case of enzymes, binding sites are designed to induce chemical changes in the molecules they bind. Hence, enzymes are the biological catalysts that manage practically every phase of the biochemistry of living systems (see Chapter 8). Cellular receptors are essentially signal-recognition units that bind physiologically active molecules, or 'chemical messengers,' such as hormones and neurotransmitters. For example, molecules of a given hormone will be transported in the blood stream to distant parts of the body. Eventually, they will come into contact with cellular receptors that are specifically shaped to interact with them. Cells possessing these receptors are often called the *targets* of the hormone. The binding of a hormone molecule induces a conformational change in the receptor protein which, in turn, triggers or switches ON a series of events in the target cell. These events may involve any phase of cell function, from activating a biochemical pathway for the breakdown of glycogen, to activating dormant genes.

targets

It is hoped that this very brief introduction to protein function will provide a basis for understanding the myriad ways in which these macromolecules serve the needs of living systems.

8

ENZYMES

In all chemical reactions a certain amount of kinetic energy, called the *energy of activation,* is necessary to get the reaction started. Activation energy is usually supplied in the laboratory by heating the *reactants* (reacting substances) together in a test tube or flask. When heat energy is transferred to molecules, their kinetic energy is increased: they move faster and hence collide harder and more often. Thus, heat increases the chemical reactivity of molecules. Chemists also often use inorganic substances known as catalysts to speed up chemical reactions, even at relatively low temperatures. The term *catalyst* describes any substance that accelerates a chemical reaction without itself undergoing any permanent chemical change.

energy of activation

reactants

catalyst

A catalyst acts, not by helping the reactants acquire activation energy, but by decreasing the amount of activation energy needed by molecules in order to take part in a chemical reaction. This principle is of great importance in biology. The metabolic activities of living cells involve a large number of biochemical reactions in which diverse substances are continually being broken down and built up. Most of the biochemical reactions that proceed so quickly and smoothly in the body would never take place at all in a test tube at physiological conditions of temperature and pH. For example, proteins, fats, and carbohydrates are normally broken down in the body to simpler substances. To accomplish this in a test tube, these substances would have to be boiled for hours in strong acid solutions. Such extremes of temperature and pH would destroy a living cell. Biological catalysts are therefore vitally necessary to lower the amount of activation energy required for the molecules in living organisms to react. These biological catalysts are called *enzymes*. Enzymes enable the cell to carry on

enzymes

its chemical activities with maximum speed and efficiency under conditions that are compatible with life.

GENERAL PROPERTIES OF ENZYMES

Most enzymes are proteins, but some are now known to be made of RNA. Up to a few years ago, it was an established belief that enzymes were exclusively proteins. However, in the early 1980's, ribonucleic acid (RNA), once thought to be a more or less passive conveyor of genetic information, was discovered to have catalytic activity.* The RNA enzymes were appropriately named *ribozymes*. As we shall see when we discuss nucleic acids further on in this text, the discovery of enzymatic RNA had a profound impact on scientific thinking about the origins of life on earth. This chapter will focus on the much more numerous and well-studied protein enzymes of living systems, bearing in mind that the basic properties and modes of action of protein enzymes are no doubt largely shared by their putative evolutionary antecedents, the ribozymes.

ribozymes

The molecules of each enzyme have a characteristic conformation. This distinctive three-dimensional shape enables enzyme molecules to recognize other molecules by their shape and to bind these molecules at specific sites. However, enzymes are highly specialized molecules that do more than bind other substances—*they catalyze chemical changes in the substances they bind*. Hence, special terms are used for enzymes, e.g., the binding site on the enzyme molecule, where the catalytic action takes place, is called the *active site*, and the ligand molecule that fits into the active site, and is chemically altered by the catalytic action of the enzyme, is called the *substrate*.

conformation

active site

substrate

Enzymes are highly specific. An *inorganic* catalyst can be used in a variety of chemical reactions, but enzymes will catalyze only one or two reactions at the most, and they exhibit a high degree of specificity with respect to the substrates they bind. An enzyme can recognize the difference between structural or optical isomers of a compound and selectively bind only one of them. Because of their specificity, a large number of different enzymes are required for the many complex biochemical reactions that can take place in the body.

Enzyme activity depends on temperature and pH factors. The most favorable conditions under which enzymes are most active

* Thomas Cech and Sidney Altman, who independently made the discovery, were awarded the Nobel Prize in Chemistry in 1989.

are called *optimum,* that is, optimum temperature and optimum pH. The optimum temperature for enzymes in the human body is body temperature (about 37°C). Optimum pH values vary to some extent: pepsin, a digestive enzyme in gastric juice, has an optimum pH of 1.5; most other enzymes in the body have pH optima ranging from 7 to 8.

optimum conditions

Enzymes can be denatured. Like other proteins, the characteristic conformation of an enzyme molecule can be changed by extremes of temperature, and pH, heavy metals, organic solvents, and concentrated salt solutions. When the native shape of an enzyme molecule is altered by *denaturation,* it ceases to function as a catalyst.

denaturation

Enzymes are not used up or permanently changed by the chemical reactions they catalyze. The enzyme molecule participates in the reaction and speeds it up, but when the reaction is done, the enzyme is chemically unchanged and can go back and repeat the whole performance over and over again. This does not mean that enzyme molecules last indefinitely; *all* biomolecules in the body are sooner or later broken down and replaced (see *turnover,* p. 103).

Some enzymes exist in two or more different molecular forms. Different species of the same enzyme are called *isozymes.* Isozymes may vary in the *rates* at which they catalyze the same chemical reaction. For example, there are five known isozymes of *lactic dehydrogenase,* an important enzyme that catalyzes an energy-producing reaction in muscles. The isozyme variant with the most rapid rate of catalysis is found in muscles that are continuously active, whereas the slower isozymes are found in muscles that are used only occasionally. Isozymes thus represent a biological adaptation to the functional requirements of different types of tissue in the body.

isozymes

Some enzymes are produced in the form of an inactive precursor—a proenzyme, or zymogen. These enzymes are mainly *proteolytic* enzymes, that is, enzymes that catalyze the hydrolysis of proteins (*proteo,* protein; *lysis,* dissolving). Such a mechanism no doubt evolved as a safety precaution; if an active proteolytic enzyme were synthesized and turned loose in a cell, it could destroy the structural and functional proteins of the cell. Zymogens are activated by other factors (often by other enzymes) at the appropriate moment when they can carry out their function without harming the organism. A good example of this is the proteolytic digestive enzymes secreted as zymogens by cells of the stomach and the pancreas into the gastrointestinal tract. When activated, their function is to catalyze the breakdown of protein molecules in the foods we eat. Another notable example is the

zymogen
proteolytic enzymes

blood coagulation — *coagulation cascade,* a sequence of complex catalyzed chemical reactions in blood that ends with the formation of a clot. This sequence is triggered by the conversion of zymogens to active proteolytic enzymes (for example, prothrombin to thrombin).

The functional activity of many enzymes requires the presence of additional nonprotein components, such as prosthetic groups, cofactors, and coenzymes. As noted previously, the protein moiety of enzymes of this type is termed the *apoenzyme*. By themselves, apoenzymes are inactive. The enzyme molecule is catalytically active only in association with its nonprotein component; the complex of the two is called the *holoenzyme*. In general, nonprotein components that are tightly bound to the enzyme molecule are called prosthetic groups, while those that are loosely bound are called cofactors (see also page 132, Chapt. 7). The terms, cofactor and coenzyme are often used interchangeably. A *coenzyme* is not necessarily specific for only one enzyme; groups of apoenzymes often operate with the same coenzyme. Many of the coenzymes in the body are derivatives of the vitamin B complex (Table 8-1). This fact explains the necessity for small amounts of these vitamins in human nutrition. B vitamins play a vital role as coenzymes in the oxidation reactions that yield energy in the cell (see Chapter 10). The metal ion cofactors of enzymes include iron, copper, zinc, and calcium ions (Table 8-2).

Margin terms: apoenzyme; holoenzyme; coenzyme

Table 8-1. Coenzymes derived from B complex vitamins

Vitamin	Coenzyme	Enzymes
Thiamine (vitamin B_1)	Thiamine pyrophosphate (TPP)	Dehydrogenases
Riboflavin (vitamin B_2)	Flavin mononucleotide (FMN) Flavin adenine dinucleotide (FAD)	Dehydrogenases
Niacin (nictotinic acid and nicotinamide)	Nicotinamide adenine dinucleotide (NAD)	Dehydrogenases
	Nicotinamide adenine dinucleotide phosphate (NADP)	
Pyridoxine (vitamin B_6)	Pyridoxal phosphate	Amino acid decarboxylase and transaminase
Pantothenic acid	Coenzyme A	Pyruvate dehydrogenases
Biotin	—	Acetyl CoA carboxylase Pyruvate carboxylase
Folic acid (and related compounds)	Tetrahydrofolic acid (THF)	Thymidylate synthetase
	Tetrahydrobiopterin	Phenylalanine hydroxylase
Cobalamin (cobalt-containing vitamin B_{12})	Coenzyme B_{12}	Methylmalonyl mutase

Table 8-2. Some metal ion cofactors of enzymes

Metal	Enzymes
Calcium (Ca^{2+})	Salivary amylase, pancreatic amylase
Copper (Cu^{2+} or Cu^{+})	Ascorbic acid oxidase
	Monoamine oxidase (MAO)
	Cytochrome oxidase
Iron (Fe^{2+} or Fe^{3+})	Catalase
	Cytochromes
Magnesium (Mg^{2+})	Phosphotransferases
Manganese (Mn^{2+})	Arginase
Molybdenum (Mo^{2+})	Xanthine oxidase
Zinc (Zn^{2+})	Carbonic anhydrase
	Carboxypeptidase

Enzymes are very powerful and efficient catalysts. Recall the reaction

$$H_2O + CO_2 \rightleftharpoons H_2CO_3$$

which represents a crucial phase in respiration (see p. 59, Chapt. 3). The two reactants combine very sluggishly *in vitro*. If the reaction had the same slow rate *in vivo,* removal of the CO_2 produced by the tissues would be hopelessly inadequate for the needs of the body. However, the reaction in red blood cells is catalyzed by the enzyme *carbonic anhydrase,* which accelerates it to such a tremendous extent that millions of molecules of CO_2 can be hydrated per minute. Furthermore, unlike the chemical catalysts used in laboratory reactions, enzyme catalysts operate with almost 100% efficiency, with no side reactions and no wastage.

Naming of Enzymes

In general, with the exception of some of the enzymes that were identified and named many years ago (such as cytochrome, pepsin, and trypsin), enzyme names end with the suffix *-ase*. The first part of the name usually indicates either the substrate acted on or the type of chemical reaction that is catalyzed. The modern system of naming enzymes is useful and informative; examples of such nomenclature are given in Table 8-3.

HOW ENZYMES ACT

We have noted previously that binding proteins have the capacity to function as receptors, carriers, antibodies, and enzymes because of their three-dimensional structure, or conformation.

Table 8-3. Naming of enzymes

Class of enzymes	Type of reactions catalyzed
Anhydrases	Removal of H_2O from substrate
Carboxylases (or decarboxylases)	Removal of CO_2 from carboxyl groups
Catalases	Breakdown of hydrogen peroxide (H_2O_2) to H_2O and O_2
Dehydrogenases	Removal of hydrogen atoms from substrates
Isomerases	Conversion of one isomer to another
Hydrolases	Hydrolysis
Ligases	Coupling together of two substrates with expenditure of energy
Lyases	Removal of groups, *not* by hydrolysis
Oxidoreductases	Oxidation-reduction reactions (electron transfers)
Oxidases	
Dehydrogenases	
Peroxidases	
Phosphorylases	Addition of inorganic phosphate groups to substrates
Polymerases	Polymerization reactions
Transferases	Transfer of chemical groups from one substance to another
Transaminases*	
Transamidases	
Phosphotransferases (kinases)	

Name of enzyme	Substrate
Amylase†	Starch (*amyl*, starch)
Arginase	Arginine
ATPase	Adenosine triphosphate (ATP)
Lactase	Lactose
Lipase	Fats (lipids)
Maltase	Maltose
Nucleases	
Deoxyribonuclease (DNAase)	Deoxyribonucleic acid (DNA)
Ribonuclease (RNAase)	Ribonucleic acid (RNA)
Nucleotidase	Nucleotides
Di- and tripeptidases	Di- and tripeptides
Aminopeptidase	Peptide bonds of terminal amino acids at free amino end of chain
Carboxypeptidase	Peptide bonds of terminal amino acids at free carboxyl end of chain
Phosphatase	Phosphate esters
Protease‡	Proteins (in general)
Sucrase§	Sucrose
Urease	Urea

* Also called aminotransferases.

† An old name for salivary amylase is "ptyalin."

‡ An old name for protease is "cathepsin."

§ An old name for sucrase is "invertin," or "invertase."

The loops and folds of the polypeptide chains of these molecules form crevices and pockets by which they recognize and bind other molecules that fit into them. The shape of the bound molecule is said to be *complementary* to the shape of the binding site of the protein. The bonds formed in these interactions are relatively weak, noncovalent bonds that are easily broken and re-formed.

complementary shape

An enzyme will temporarily bind one or more specific substrates to form an *enzyme-substrate complex*. A chemical reaction is then catalyzed in the active site of the enzyme that changes the substrate to one or more *end products*. The end product then breaks off from the complex, leaving the enzyme molecule unchanged and ready to pick up another substrate molecule. This process can be represented as follows:

enzyme-substrate complex

end products

$$\underset{\text{(substrate)}}{S} + \underset{\text{(enzyme)}}{E} \rightleftharpoons \underset{\text{(enzyme-substrate complex)}}{ES} \rightleftharpoons \underset{\text{(end product)}}{P} + \underset{\text{(enzyme)}}{E}$$

As in the case of other binding proteins, the lock-and-key model is relevant to enzyme-substrate interactions. Enzymes have a specially shaped active site, into which only specific substrates or portions of substrate molecules can fit. The enzyme may thus be considered the "lock" and the substrate, the "key." If the key fits in the lock, it can be turned and the door will open; that is, the chemical reaction will proceed (Fig. 8-1).

The rate of enzyme action is regulated by the concentrations of the substrate, the enzyme, and the end products. The rate of a given reaction increases to a maximum when large amounts of substrate and enzyme molecules are available and tends to slow down as the end products accumulate. Some reversible reactions are catalyzed in both directions by an enzyme when the accumulated end products combine with the enzyme and are converted back to the original substrate. An example of this type of reaction is

$$H_2O + CO_2 \underset{\text{anhydrase}}{\overset{\text{carbonic}}{\rightleftharpoons}} H_2CO_3$$

In the tissues, where CO_2 diffuses out of cells into the blood stream (and into red blood cells), the enzyme carbonic anhydrase accelerates the reaction to the right. In the lungs, where CO_2 diffuses out of the blood stream, and is blown off by the lungs in expired air, the equilibrium of the reaction shifts to the left. The enzyme then accelerates the reverse reaction.

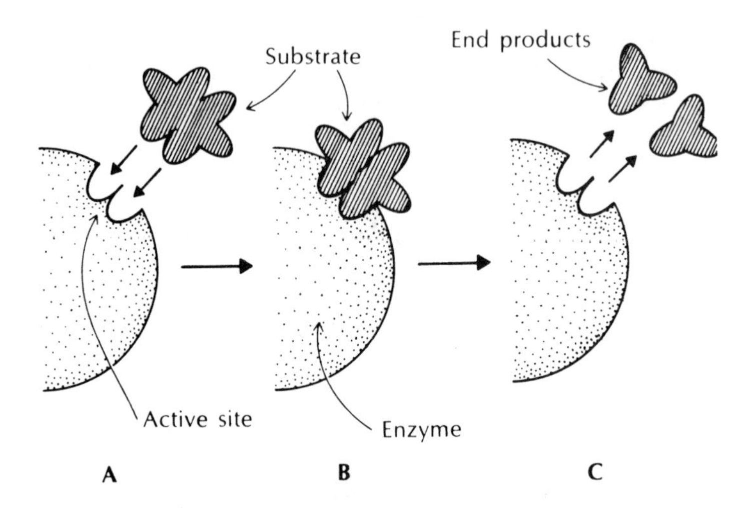

Fig. 8-1. Lock-and-key model of enzyme action. **A,** The substrate molecule fits into the active site on the surface of the enzyme molecule. **B,** An enzyme-substrate complex is formed; and a chemical reaction is catalyzed, in which the substrate is changed. **C,** The end product(s) of the reaction dissociate from the unchanged enzyme molecule. The diagram illustrates only one of many possible modes of enzyme action, namely, the splitting of a molecule of substrate into two smaller molecules. This reaction is shown as occurring in the breakdown direction only, but the reverse may be catalyzed as well in certain types of reactions. If the two end products attached to the active site properly, they could become the substrates and the substrate shown above would be the end product.

However, there are chemical reactions in the body where the energy gradient for the reaction to the right, e.g., substrate → product, is more favorable than for the reverse reaction, product → substrate. These reactions are essentially *irreversible* because enzymes cannot force reactants to move 'uphill' against an energy gradient. In such circumstances, one enzyme will catalyze the more favorable 'downhill' reaction, while the reverse 'uphill' conversion of product to substrate will be coupled to an energy-yielding reaction, and catalyzed by a different enzyme. A good example of this is the overall reaction in which stored glycogen is broken down to yield glucose (by the process of *glycogenolysis*), and its reverse, the biosynthesis of glycogen from glucose (by the process of *glycogenesis*). Both processes take place in liver cells (see also p. 94 in Chapt. 5), and may be symbolized as follows:

glycogen → (many) glucose

glycogen ← (many) glucose

The first reaction, the breakdown of glycogen, is energetically 'downhill,' and is catalyzed by one set of enzymes. The second reaction, the synthesis of glycogen, is not simply a reversal of the first. It is energetically 'uphill,' and involves a different series of reactions, and a different set of enzymes.

FUNCTION OF ENZYMES

All living organisms are coordinated chemical machines that run on the catalytic power supplied by enzymes. The organism itself synthesizes the enzymes it needs according to a genetic program handed down from generation to generation. Moreover, the patterns of enzyme synthesis and enzyme action are the same whether they occur in the most primitive bacterial cell or in the human cell. It should be evident by now that enzymes are a fundamental part of human function and that knowing something about them is essential to an understanding of human physiology in health and disease. Although enzymology is a complex science, a good introduction to the subject is provided by the digestive enzymes. The substrates here are food molecules we are all more or less familiar with, and the mechanisms of chemical digestion are not particularly difficult to follow. The enzymology of digestion is dealt with later on in this chapter. Some important aspects of enzymes as functional biomolecules are considered here.

Regulatory (Allosteric) Enzymes

In complex organisms, such as humans, enzymes are found in extracellular fluids (such as blood) and in various cellular secretions (digestive juices, tears, semen). The great majority of enzymes are found in cells and on cell membranes (cellular enzymes). Regardless of where they are found, enzymes are all synthesized *inside* cells. The chemical reactions they catalyze are often steps in a sequence; that is, substrates are changed—broken down (degraded) or built up (synthesized)—not all at once, but gradually. In most cases, each step in the sequence is catalyzed by a different enzyme. An array, or assembly, of enzymes linked with specific substrates in a correlated series of reactions is called a *metabolic pathway;* the chemical substances produced in such pathways are *metabolites*.

metabolic pathway
metabolites

We shall consider a hypothetical pathway in which one substrate (S_1) is converted to an end product (P_1) by the catalytic action of an enzyme (E_1). The end product of the first reaction then becomes the substrate (S_2) of the second reaction, catalyzed by the second enzyme (E_2), and so on. This can be represented as follows:

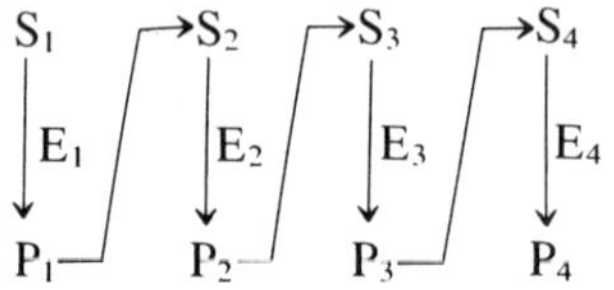

S_1 $\xrightarrow{E_1}$ P_1 → S_2 $\xrightarrow{E_2}$ P_2 → S_3 $\xrightarrow{E_3}$ P_3 → S_4 $\xrightarrow{E_4}$ P_4

feedback inhibition

regulatory enzyme

allosterism

When a pathway of this type is analyzed, it is often found that the activity of the first enzyme in the series (E_1) appears to be inhibited as the last metabolite (P_4) accumulates; or more briefly, P_4 can switch off E_1. Thus, the end product of the pathway *regulates its own synthesis*. This phenomenon is called (negative) *feedback inhibition;* it is a prime example of self-regulation, the off-on switches that are built into biomolecular systems. The key enzyme in the pathway that is inhibited in this manner (enzyme E_1 in the above pathway) is termed a *regulatory enzyme*. Enzyme molecules of this type tend to be larger than others and have an active site *plus one or more additional binding sites* that are situated on other regions of the molecule. When a protein has sites for binding more than one ligand, it is said to exhibit the property of *allosterism* (*allo,* other; *stereos,* three-dimensional shape). If the protein is an enzyme, it is called an allosteric enzyme. Allosteric regulatory enzymes have a dual function; they are specific catalysts, and they switch metabolic pathways off (or on) in response to the chemical signals they receive. How does a small molecular species like our hypothetical metabolite P_4 (which is often called an effector, modulator, or modifier) inhibit the regulatory enzyme? It is believed that when the effector occupies the allosteric binding site on the enzyme, it changes the conformation of at least some portion of the enzyme molecule. The change in shape somehow decreases the affinity (chemical attraction) of the active site for its customary substrate.

The feedback is illustrated as follows with the regulatory enzyme (E_1), its substrate (S_1), the active site (a), and the binding site (b):

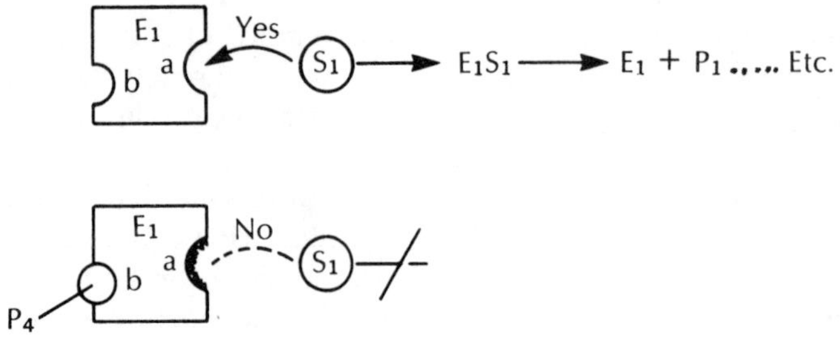

E_1 (b, a) ← Yes — S_1 → E_1S_1 → $E_1 + P_1$ ..... Etc.

P_4 — E_1 (b, a) ← No - - - S_1 ⟶/

When the binding site is empty, the enzyme-substrate interaction takes place; when the binding site is occupied by the effector molecule (P_4), the enzyme-substrate interaction is inhibited. It should be noted that although the effector is bound to the enzyme, it takes no part in the reaction the enzyme catalyzes, and the noncovalent bonds it forms with the enzyme are completely reversible.

In some instances, the affinity for a substrate is *increased,* rather than decreased, when an allosteric binding site is occupied. The binding of oxygen by the hemoglobin molecule is a classic example of this sort of effect. Hemoglobin is *not* an enzyme, but it is an allosteric protein. One of its functions is to carry oxygen, which it loosely binds in the lungs and unloads in the tissues. The hemoglobin molecule has four distinct iron-containing binding sites, each one of which binds 1 oxygen (O_2) molecule. When the first O_2 molecule is bound, it greatly increases the affinity of the other three binding sites for O_2. When the second molecule is bound, it becomes even easier for the remaining two sites to bind oxygen. The fact that oxygen can be *cooperatively* bound at allosteric sites on the hemoglobin molecule greatly contributes to the effectiveness of hemoglobin in carrying out its major function, oxygen transport.

Competitive Inhibition

The catalytic activity of an enzyme may be showed down when another molecule that resembles the normal substrate competes with the substrate for binding to the active site. If the competitive molecule, or *inhibitor* (I), occupies the active site, it temporarily prevents the formation of an enzyme-substrate complex:

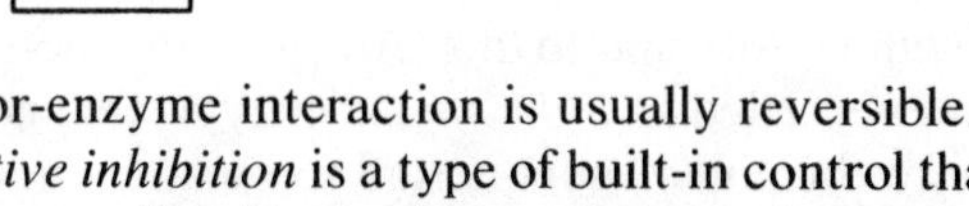

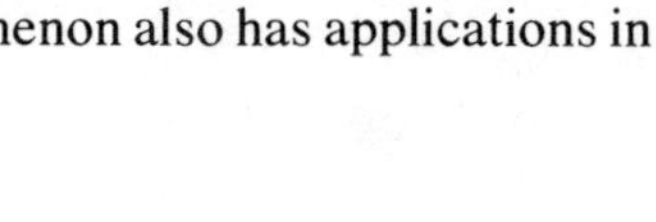

The inhibitor-enzyme interaction is usually reversible.

competitive inhibition

Competitive inhibition is a type of built-in control that operates in enzymes, as well as in membrane carrier and receptor proteins. The phenomenon also has applications in *chemotherapy,* a branch

chemotherapy

of medicine that deals with the treatment of disease by chemical agents. In recent years, with advances in the knowledge of enzymes and in chemical technology, chemists have synthesized a variety of inhibitors that effectively block specific enzymes. For example, microorganisms (including viruses) or cancer cells have an array of enzymes that make it possible for them to survive and flourish in the body. If one or more of these enzymes are competitively inhibited by chemotherapeutic agents, such as certain antibiotics or antimetabolites (used in the treatment of cancer), the growth and function of the invaders are impaired to the extent that they become easy targets for destruction by the body's immune mechanisms.

curare

The paralytic effects of the Indian arrow poison, *curare,* provide a very interesting example of how two substances may compete for a binding site on a protein molecule. The protein involved here is not an enzyme; it is a receptor protein on the membrane of skeletal muscle cells that has a special binding site for the neurotransmitter acetylcholine. When the nerve ending on the muscle cell becomes excited, it releases acetylcholine, which binds to the receptor, forming a receptor-acetylcholine complex. This event triggers the contraction of the muscle cell. Curare and acetylcholine have relatively similar structures, and curare can occupy the same binding sites. However, although the curare "key" readily "fits" into the lock, it does not "turn." The receptor-curare complex cannot initiate muscle contraction. On the contrary, by blocking the normal binding of acetylcholine, curare paralyzes the muscle. The binding of curare is reversible if the administration of the drug is carefully controlled. Derivatives of curare are routinely used to induce skeletal muscle relaxation during surgery.

Enzyme Induction

enzyme induction

Perhaps the most dynamic aspect of enzymes in living systems is the phenomenon of *enzyme induction.* Recall that the biosynthesis of enzymes is genetically programmed; it proceeds according to encoded 'instructions' in genes. The genes that direct the synthesis of certain enzymes in cells can be *activated* (switched ON) or *repressed* (switched OFF) in response to various extracellular or intracellular signals, such as nutrients, metal ions, hormones, growth factors and toxins. An agent that activates a gene or set of genes is called an *inducer;* the enzyme that is synthesized in response to the given agent is said to be an *inducible enzyme.* In Chapter 12, we shall discuss the basic model for enzyme induction. It is an assembly of interrelated genes, collectively termed

inducer
inducible enzyme

an *operon*. ON/OFF genetic controls of enzyme synthesis enable the cell to adapt with maximum efficiency to changes in its external and internal environment.

operon

Enzyme-Hormone Interactions

A *hormone* may be regarded as a chemical messenger, sent out by one set of cells in the body, to program and regulate the activity of other cells *(target cells)* that are responsive to the hormone. Although they have a wide range of physiological effects, essentially they all operate by modifying certain metabolic pathways in target cells. Their effects are thus mainly mediated by cellular enzymes. An excellent model for hormone action was suggested in the 1940's by the discovery of a *second messenger,* the substance cyclic adenosine monophosphate (cyclic AMP) (see Chapter 11). In this model, the hormone molecule (the *first* messenger) does not enter the cell. It is recognized by a specific receptor protein on the membrane of the target cell and fits into its binding site. The hormone-receptor complex has an altered conformation that somehow activates the enzyme *adenylate cyclase,* which is situated on the inner aspect of the cell membrane. The activated enzyme immediately catalyzes a reaction inside the cell that converts adenosine triphosphate (ATP) to cyclic AMP, the *second* messenger. Cyclic AMP in turn activates a key regulatory enzyme in the cell; in effect, this switches on a green light that sets certain metabolic pathways in motion. The cascade of effects can be summarized as follows:

hormone

target cells

second messenger

adenylate cyclase

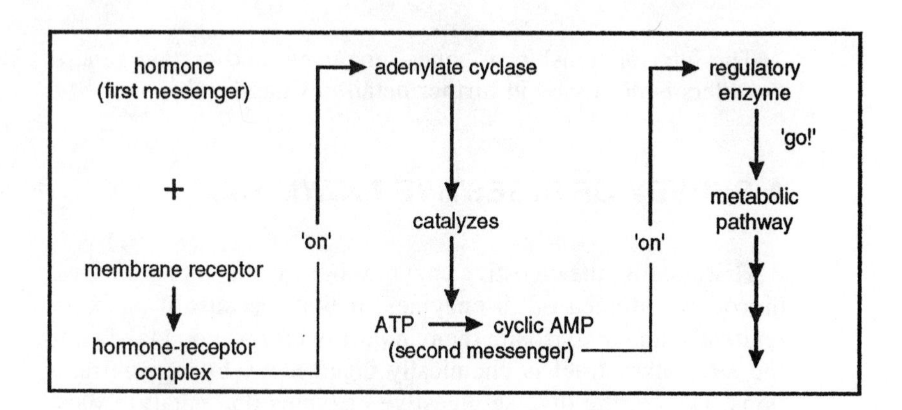

In this manner, by a defined sequence of signals, hormones that never even enter the cell control the pattern of activity in a target cell. The second messenger model appears to be valid mainly for the nonsteroid hormones. The steroid hormones, which are lipid soluble, penetrate the cell membrane and are carried into the nucleus. In some instances, they are known to turn on genes to *induce* certain cellular enzymes.

Enzyme Defects and Genetic Disease

1. Most enzymes are proteins that are synthesized in the cell according to the instructions of an inherited genetic program.
2. All living organisms require enzymes to carry on the chemical processes necessary for survival, growth, and reproduction (the sum total of these processes is *metabolism*).

If we take into account these two fundamental facts about enzymes, it follows that faulty inherited factors (genes) may result in absent or defective enzymes and that this in turn may produce serious genetic disorders, termed *inborn errors of metabolism:*

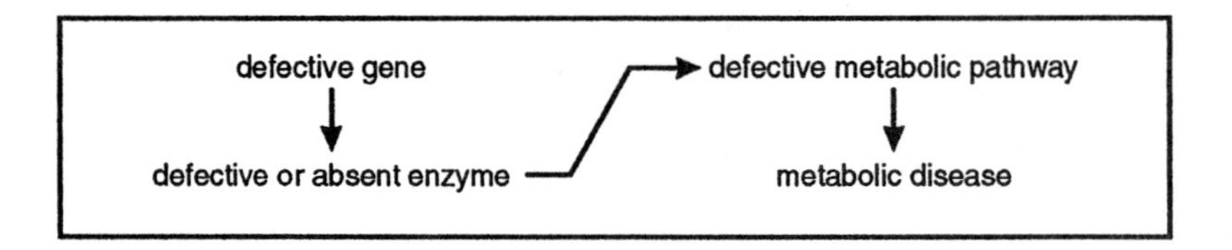

The interrelationship of genes, enzymes, and certain genetic disorders is discussed in further detail in Chapter 13.

A SURVEY OF DIGESTIVE ENZYMES

hydrolases

Historically, the digestive enzymes have been among the most thoroughly studied of all enzymes, mainly because it has been relatively simple to isolate them in purified form. Food taken into the alimentary tract is chemically digested by hydrolysis reactions. Hence, the diverse digestive enzymes that catalyze these reactions are generally classified as *hydrolases*. The hydrolysis of large complex food molecules, such as proteins, starches, fats, and nucleic acids, to smaller molecules is necessary because only the smaller molecules can be absorbed from the digestive tract. In the process of chemical digestion, each of these classes of large food molecules is successively split up into smaller and smaller sized fragments in a stepwise series of reactions, as follows:

1. Carbohydrates (starches and complex sugars) are gradually hydrolyzed to monosaccharides (simple sugars).
2. Proteins are gradually hydrolyzed to amino acids.
3. Fats are gradually hydrolyzed to glycerol and fatty acids.
4. Nucleic acids are gradually hydrolyzed to their component nitrogenous bases, pentose sugars, and phosphoric acid.

Most of the digestive enzymes are present in the following *digestive juices* secreted into the alimentary tract:

1. *Saliva* (pH 6.2 to 7.4), secreted by the salivary glands into the mouth. Its main enzyme is *salivary amylase* (formerly called ptyalin).

salivary amylase

2. *Gastric juice* (pH 1.6 to 2, due to the presence of hydrochloric acid, HCl), secreted by mucosal cells of the stomach. It contains the enzymes *pepsin* and *gastric lipase*.

pepsin

3. *Pancreatic juice* (pH about 8), secreted by the exocrine glands of the pancreas by way of the pancreatic duct into the small intestine. Its enzymes include *trypsin, chymotrypsin, elastase, carboxypeptidase, lipase, amylase*, and *nucleases*.

trypsin

It may be noted here that bile, which is secreted by the liver, is also a digestive juice. Bile contains *no* digestive enzymes but, as we have seen, its content of bile salts plays a vital role in the emulsification and absorption of ingested lipids of all types (see pp. 109–110, Chapt. 6).

Some digestive enzymes are not secreted in digestive juices. The enzyme molecules are membrane-bound, that is, they are embedded in the cell membranes of the simple columnar epithelium *(enterocytes)* lining the small intestine. Since these cells have large numbers of microvilli, commonly termed a 'brush border,' on their luminal surfaces, this group of digestive enzymes is called the *brush border enzymes*. They include several disaccharidases, peptidases, nucleotidases, nucleosidases, and enteropeptidase (also called enterokinase).

brush border enzymes

Tables 8-4 to 8-7 summarize the digestive enzymes, their sources, substrates, and end products.

Zymogens (Inactive Pro-Enzymes)

zymogen

As mentioned previously, most enzymes involved in the digestion of dietary proteins are not secreted in an active state but in the form of *inactive precursors,* or *zymogens.* The inactive zymogen must be converted to the active enzyme by another factor. A list of the major zymogens found in gastric and pancreatic juice, and the factors that convert them to active enzymes, is given in Table 8-8.

Table 8-4. Major enzymes involved in carbohydrate digestion (starches and sugars)

Enzyme	Source	Substrates	End products
Salivary amylase (ptyalin)	Saliva	Starch (polysaccharides)	Intermediate-sized polysaccharides (dextrins) Some maltose
Pancreatic amylase	Pancreatic juice	Starch Dextrins	Maltose
Disaccharidases	Brush border of intestinal cells	Disaccharides	Monosaccharides
Maltase		Maltose	Glucose
Sucrase		Sucrose	Glucose, fructose
Lactase		Lactose	Glucose, galactose

Table 8-5. Major enzymes involved in the digestion of fats

Enzyme	Source	Substrates	End products
Gastric lipase*	Gastric juice	Triacylglycerols with short chain fatty acids	Mono- & diacylglycerols Fatty acids
Pancreatic lipase§	Pancreatic juice	Triacylglycerols emulsified by *bile salts*	Monoglycerides Fatty acids Glycerol

* The activity of gastric lipase is limited by the low pH of gastric juice. It may hydrolyze triacylglycerols containing short-chain fatty acids.

§ A protein called **colipase** is also secreted in pancreatic juice. It binds to the surface of fat droplets in the presence of bile salts, and provides an attachment site for lipase.

Note: A lingual lipase is also secreted into saliva by glands of the tongue and soft palate.

Table 8-6. Major enzymes involved in the digestion of proteins

Enzyme	Source	Substrates	End products
Pepsin*	Gastric juice	Proteins	Larger & smaller polypeptide fragments
Trypsin* Chymotrypsin* Elastase* Carboxypeptidases*	Pancreatic juice	Proteins & partially digested polypeptides	Tri- & dipeptides Amino acids
Aminopeptidases Tri- & dipeptidases	Brush border of intestinal cells	Polypeptides Tripeptides Dipeptides	Terminal amino acids Amino acids Amino acids

* These proteolytic enzymes are all secreted in the form of **zymogens** (see p. 141 and Table 8-8).

Table 8-7. Enzymes involved in the digestion of nucleic acids

Enzyme	Source	Substrates	End products
Nucleases	Pancreatic juice	DNA and RNA (polynucleotides)	Nucleotides
Nucleotidases	Brush border of intestinal cells	Nucleotides	Nucleosides
Nucleosidases		Nucleosides	Purines Pyrimidines Pentose sugars

Table 8-8. Conversion of zymogens to active digestive enzymes

Source	Zymogen	Activating factor	Active enzyme
Gastric juice	Pepsinogen	HCl (in gastric juice)	Pepsin
Pancreatic juice	Trypsinogen	Enterokinase* (a brush border enzyme)	Trypsin
	Chymotrypsinogen	Trypsin	Chymotrypsin
	Proelastase	Trypsin	Elastase
	Procarboxypeptidases	Trypsin	Carboxypeptidases

* A more recent name for this enzyme is **enteropeptidase**.

9

EXCHANGES BETWEEN CELLS AND THEIR ENVIRONMENT

The first life forms undoubtedly arose in an ancient sea, and the aqueous (watery) environment it provided has persisted in living systems ever since. Sea water surrounded our primitive unicellular (one-celled) ancestors; as more complex, multicellular (many-celled) creatures evolved, this external aqueous medium became enclosed within them in the form of extracellular fluid (fluid *outside* cells), which still bears a close chemical resemblance to the seas that bathed the earliest living cells. The fluid of the body is conventionally allotted to *compartments,* separated from each other by biological barriers *(membranes).* The two major fluid compartments are *intracellular fluid (ICF),* which is the fluid *inside* cells collectively, and *extracellular fluid (ECF).* The barrier between these two compartments is the cell membrane. The ECF compartment is further subdivided into vascular fluid (blood, or more accurately, *blood plasma*) and *interstitial* (tissue) *fluid (ISF).* The barrier between these two compartments is the capillary wall.

fluid compartments
membranes
ICF
ECF
blood plasma
ISF

This chapter deals with the nature of these fluids, the barriers that separate them, and the dynamics of the exchanges that take place across the barriers.

CELLULAR AND EXTRACELLULAR SOLUTIONS

The fluids inside cells and all the body fluids outside cells are essentially aqueous solutions. *Solutions* are mixtures of two or more substances: the *solvent* (the fluid in which other substances dissolve) and *solutes* (the substances that are dissolved). Water

solvent
solutes

is the unique solvent in all forms of life on earth. It makes up more than 50% of the tissues and fluids of the body. The solvent properties of water enable it to hold in solution a large variety of solutes ranging from very small ions to very large macromolecules. The characteristics of the mixtures vary according to the size of the particles mixed with the solvent, water.

True Solutions

A true solution is a homogeneous (uniform) mixture of water and one or more solutes that are very small molecules or particles less than 1 millionth of a millimeter in size. In a true solution, the solutes are so completely dispersed in the solvent that they are invisible. For example, a glass of water with salt or sugar dissolved in it and a glass of pure water look exactly alike. A true solution may be colored, depending on the solute, but it will still be as clear as pure water. The molecules of many electrolytes break up into even smaller particles, or ions, when they are in solution in water. The solutes that form true solutions in water are called *crystalloids* because they aggregate as crystals when the water in the mixture is evaporated off.

crystalloids

Colloidal Solutions

Body fluids also contain very large molecules (macromolecules), such as proteins and polysaccharides. Macromolecules are large only in a relative sense. The biological macromolecules that form colloidal solutions in water range from just over 1 to 500 nm in diameter, that is, from over 1 millionth of a millimeter to 1/2,000 mm. The word *colloid* is derived from a Greek word meaning "resembling glue." Aqueous colloids often form viscous (thick and gluey) solutions, for example, egg albumin. Although colloids are much larger than the crystalloid particles found in true solutions, they do not settle out but remain uniformly dispersed in the solvent. One reason for this is the constant agitation of the colloidal particles produced by the collision of water molecules with the colloid molecules. Secondly, colloidal particles in a solution tend to repel each other, because colloidal particles formed by one particular type of macromolecule, for example, proteins, all carry the same electrical charge, either negative or positive. It is a physical law that particles carrying *like* charges repel each other (and opposite charges attract). This mutual repulsion keeps colloidal particles from sticking together and forming large aggre-

colloid

gates that would be heavy enough to settle out of the solvent and form sediment on the bottom.

The relatively large size of colloids, as compared with crystalloids, is responsible for another property of colloidal solutions, namely the *Tyndall effect*. When a beam of light is passed through a colloidal solution, the light is reflected and scattered from each of the individual large colloidal molecules in the solution. This effect does not occur in true solutions, where the individual solute particles are too small to reflect light.

Tyndall effect

Suspensions

In addition to crystalloids and colloids, the fluids inside and outside the cell contain suspended material much larger than any molecule. This includes, for example, the many granules suspended in the cytoplasm of cells or the suspension of red blood cells in the fluid portion of the blood. Suspensions are heterogeneous (not uniform) mixtures of water and particles that may be large enough to be visible to the naked eye. The best example of a suspension is a mixture of sand and water. The solvent and the suspended material do not blend together, because the particles are large enough to settle to the bottom unless the mixture is constantly agitated. Ordinary filter paper will stop particles suspended in water but will allow solutions of crystalloids and colloids to pass through.

THE CELL MEMBRANE

Compartmentation, that is, the subdivision of the body of an organism into specialized areas (compartments), is a consistent feature of living systems. The composition and properties of a compartment are controlled by the nature of the barrier that encloses it and separates it from other areas. The cell membrane, also called the *plasma membrane,* is the essential barrier at the surface of cells. It physically defines the boundaries of the cell as a discrete compartment and, by controlling the movement of materials into and out of the cell, it allows the cell to maintain a characteristic internal environment. It may be noted here that similar membrane barriers envelope various organelles within the cell, creating functionally specialized subcompartments. The surface membrane and the subcellular membranes are not identical, but they have a common architecture and similar properties. The much-studied cell membrane is considered the basic model for all cellular membranes.

compartmentation

plasma membrane

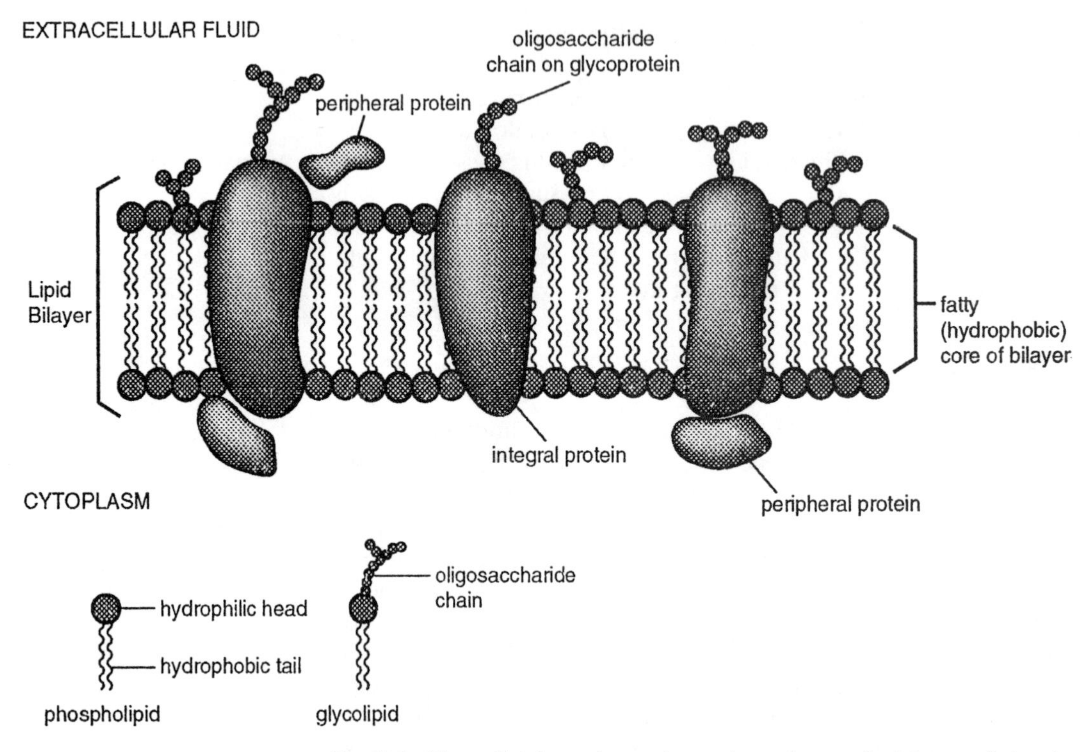

Fig. 9-1. The cell (plasma) membrane is made up of a bilayer of closely packed lipid molecules. The hydrophilic heads of the lipids face the aqueous extracellular or intracellular (cytoplasmic) fluids, and their hydrophobic (fatty hydrocarbon) tails are sequestered in the interior core of the bilayer. Amphipathic integral proteins are embedded in the bilayer; peripheral proteins are associated with the membrane surfaces. Oligosaccharide (carbohydrate) groups attached to lipids and proteins are always on the external face of the membrane.

The membranes that enclose mammalian cells are delicate fluid films ranging from about 7–10 nm in width. They are made up of an assembly of lipid and protein molecules in association with smaller amounts of carbohydrates. The carbohydrates are in the form of oligosaccharide chains which are attached to some lipids (glycolipids) and proteins (glycoproteins). Fig. 9-1 is a schematic diagram showing the characteristic architecture and major molecular components of the cell membrane.

The Lipid Bilayer

The structural matrix of the cell membrane is a bimolecular sheet of lipid molecules, or a *lipid bilayer,* composed of phospholipids, cholesterol and glycolipids. Phospholipid and cholesterol molecules are the most numerous, and are found in approximately equal quantities in mammalian cell membranes; glycolipids are the least abundant. Recall from Chapter 6 that phospholipids and glycolipids are amphipathic molecules; each has a polar hydrophilic head and a fatty hydrophobic tail. We have noted that when these amphipathic lipids are placed in aqueous media, such as the fluids inside and outside the cell, they spontaneously assemble themselves into a bilayer (refer to Fig. 6-4). The key organizing forces in the formation of the bilayer are *hydrophobic interactions.*

lipid bilayer

hydrophobic interactions

As shown in Fig. 9-1, the predominant phospholipid molecules are arranged in two single rows (monolayers or leaflets) with their hydrophilic heads facing the extracellular and intracellular fluids, and their hydrophobic tails aggregated in the fatty interior core of the bilayer, avoiding contact with water. The small number of glycolipids in the membrane have the same orientation as the phospholipids, but are found *only in the outer monolayer.* Not shown in the diagram are the mainly hydrophobic cholesterol molecules which are located in the interior of the bilayer, usually near the heads of the phospholipids. The amount of cholesterol in the membrane largely determines its fluidity. In general, membranes containing more cholesterol are less fluid.

Because of the hydrophobic effect, lipid bilayers are typically rounded-off structures that have no edges. Furthermore, they are *self-sealing,* that is, when the continuity of the bilayer sheet is broken and the hydrophobic groups at the edges are exposed to the surrounding aqueous media, the bilayer spontaneously closes up and seals itself at the site of the break. As we shall see, a common activity of the surface cell membrane, as well as the membranes that enclose organelles in the interior of the cell, is budding or pinching off small fragments of themselves. These fragments instantly shape up into small sealed-off spherical bodies called *vesicles* (derived from the Latin term *vesicula,* meaning a small fluid-filled sac). The hydrophobic interactions that result in the formation of vesicles are also responsible for sealing up the minute gaps left in the parent membranes by this process.

self-sealing

vesicles

Membrane Proteins

integral proteins
peripheral proteins

transmembrane proteins

The proteins of cell membranes are classified as *integral proteins* or *peripheral proteins* according to whether or not they penetrate the lipid bilayer. Integral proteins are embedded in the bilayer. The majority of them are *transmembrane proteins* which not only span the full width of the bilayer, but also protrude for varying distances into the fluids on both sides of the membrane. Most of the integral proteins of mammalian cell membranes that have thus far been studied have been found to be glycoproteins. The oligosaccharide chains of these proteins, like those of the membrane glycolipids, are always attached to the portion of the molecule that faces the external (extracellular) medium. Peripheral proteins are exclusively surface molecules that do not extend into the bilayer. They are usually bound by weak noncovalent forces to the protruding portions of integral proteins or lipid polar heads on the inner and outer surfaces of the membrane.

Like the lipids of the membrane, the membrane proteins are amphipathic molecules. In the case of the integral proteins, the surfaces of the molecules that are immersed in the fatty core of the bilayer consist of hydrophobic regions, whereas hydrophilic groups predominate on the surfaces of the molecules that protrude out of the bilayer into the intracellular and extracellular fluids. Some transmembrane proteins provide a passageway for water-soluble ions and molecules to cross the lipid bilayer. Protein molecules of this type have an added structural feature, namely, an internal aqueous (hydrophilic) pore running through them that communicates with the fluids on both sides of the membrane. Peripheral proteins have a somewhat different arrangement of their hydrophobic and hydrophilic regions. Since they are surrounded by aqueous media, their exposed surfaces are hydrophilic, and their hydrophobic groups are segregated in the interior of the molecules. In general, peripheral proteins are water-soluble proteins that can be removed from the membrane with relative ease. In contrast, the integral proteins can be displaced from the membrane only by disrupting the bilayer structure with detergents or organic solvents.

The proteins of the cell membrane account for most of its complex functional activities. They serve as channels, ion pumps, carriers, enzymes, receptors, and cell surface antigens (or identity tags).

A Summary of Cell Membrane Properties

Cell membranes are fluid structures. A major aspect of cell membranes is their fluidity and the mobility of membrane components. Each monolayer may be considered a two-dimensional fluid film in which the lipid and protein molecules are free to diffuse laterally (sideways) through the plane of the membrane. The measured rates of lateral diffusion of membrane components indicate that cell membranes normally have a fluid consistency resembling that of an oil.

Cell membranes are asymmetrical structures. All biological membranes that have been studied thus far have been found to have an asymmetrical distribution of components. In other words, cell membranes exhibit a characteristic 'sidedness' in that the outer and inner monolayers are different. To begin with, different kinds of phospholipids are present in the two layers of the membrane. Secondly, integral proteins span the membrane asymmetrically, and each particular species of integral protein is always oriented in the same direction. Thirdly, as we have seen, the oligosaccharide groups of membrane glycolipids and glycoproteins *always face the extracellular side of the membrane.* Indeed, cells have a 'sugar coating'—a feltlike mass of oligosaccharide chains termed a *glycocalyx*—extending from the external surface of the membrane. The asymmetry of the membrane has functional significance. For example, a membrane protein that is a receptor for a hormone or neurotransmitter would have to have its binding site on the external face of the membrane in order to interact with these chemical signals; otherwise it would be totally ineffective as a receptor. Similarly, the external orientation of oligosaccharide groups on the membrane enables them to serve as recognition sites in intercellular communication systems.

glycocalyx

It is important to bear in mind that the asymmetry of the cell membranes is always maintained. Firstly, although lateral movement of membrane components is common, movement from one monolayer to the other, which is termed *flip-flop,* is severely restricted. Flip-flop movement is energetically unfavorable because hydrophilic groups on membrane components would have to be forced to traverse the fatty hydrophobic core of the bilayer in order to change their positions. Secondly, the rule of asymmetry is strictly preserved even though membrane constituents undergo rapid turnover cycles, and fragments of membrane (in the form of vesicles) are more or less constantly being pinched off or incorporated into the membrane structure.

permeability

The cell membrane is a selectively permeable barrier. The major functional property of a biological barrier is its *permeability,* that is, the degree to which it allows substances to pass through (permeate) it. Cell membranes are highly selective barriers. The fatty hydrophobic core of the membrane serves as a barrier to the movement of hydrophilic substances, namely, ions and polar molecules. In maintaining the integrity of the intracellular environment in changeable and sometimes hostile surroundings, this permeability barrier is essential to cell survival. At the same time, the cell must have ready access to nutrients and other substances from the external environment, and it must also be able to release various substances, such as secretory products and waste metabolites, to the outside. It follows that the cell membrane is not just an inert physical barrier, but a vital structure across which there is a continuous and highly organized traffic of substances into and out of the cell. The selective permeability of the cell membrane reflects both the barrier properties of its lipid matrix, and the functional activities of its range of specialized protein components.

PASSIVE MOVEMENT OF SUBSTANCES THROUGH MEMBRANES

Diffusion

kinetic energy

Chemical substances are made up of particles that are always in motion due to their *kinetic energy*. In this context, we shall use the term 'particle' to mean the component *atoms, ions,* or *molecules* of any given substance. Some basic rules are [1] particles move faster at higher temperatures, and [2] lighter (smaller) particles move faster than heavier (larger) particles. The movement is completely random, causing the particles to collide constantly with each other and rebound. When particles are crowded together in a confined space (high concentration), the collisions and rebounds occur more frequently. As a result, the particles tend to spread out in a direction where there are fewer collisions, in other words, toward a region where they are less concentrated.

diffusion

The overall movement of particles from a region of high concentration to a region of lower concentration is called *diffusion* (meaning "to spread out"). It is said to be *passive* because the movement is 'downhill', i.e., it does not require an investment of energy. Given enough time, particles will spread out and distribute themselves uniformly in the space available to them. At that point, although they continue to move in a random manner, there can

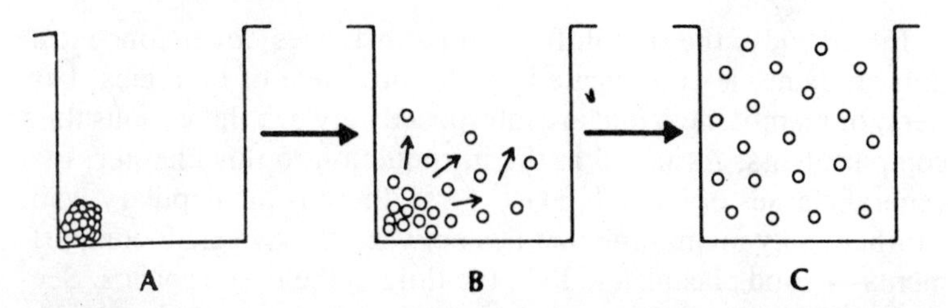

Fig. 9-2. Diffusion from an area of higher concentration to an area of lower concentration. **A,** Particles concentrated in lower left region. **B,** Through random movement in all directions, there is a net diffusion toward a region of lower concentration. **C,** Diffusion equilibrium—the particles are evenly distributed in the space available to them.

be no further net change in their distribution; once uniformly distributed, there will be equal numbers of particles moving into and out of any given region. This condition is known as *diffusion equilibrium* (Fig. 9-2).

diffusion equilibrium

Where there is a difference in the concentration of particles in two regions, there is said to be a *concentration gradient*. A gradient may be considered as a slope descending from an area of higher concentration to an area of lower concentration (Fig. 9-3). Particles diffuse passively *down* a concentration gradient in much the same way that pebbles roll down a hill.

concentration gradient

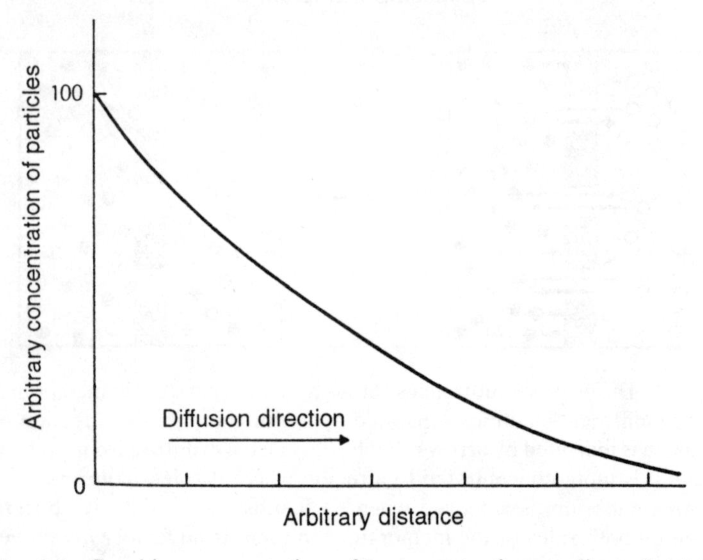

Fig. 9-3. Graphic representation of a concentration gradient. Particles will diffuse to the right from any point along the horizontal axis.

In the body, the free diffusion of substances down concentration gradients is complicated by the presence of biological barriers, or membranes, that are interposed between the various fluid compartments. As noted in the introduction to this chapter, two general classes of barriers exist. First, there is the capillary wall, a rather leaky membrane that lies between the two ECF compartments—blood plasma and ISF, the fluid in the tissue spaces. Second, there is the cell membrane (or plasma membrane), a highly selective barrier that insulates the cellular contents (ICF) from the tissue fluid (ISF) bathing the outside of cells. If a membrane separating two different solutions is *freely* permeable to all the substances on both sides of the membrane, then these substances will diffuse down their concentration gradients from one compartment to the other. Eventually there will be equal concentrations of all the substances on both sides of the membrane, and the random movement of particles will produce no further net change (Fig. 9-4). Obviously, if the membrane barrier is *impermeable* to one or another substance, those substances will not diffuse even though there may be a steep concentration gradient for them.

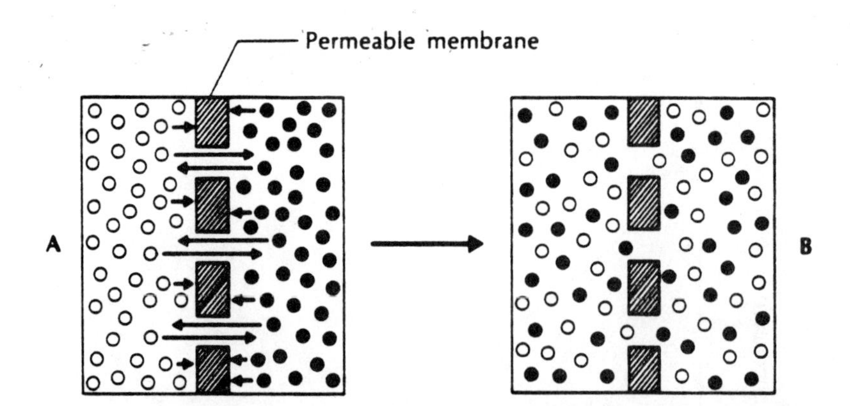

Fig. 9-4. Diffusion of substances through a freely permeable membrane. **A,** Two different solutions separated by the membrane; the direction of diffusion is indicated by arrows. Each substance will diffuse from a region where it is more concentrated to a region where it is less concentrated. **B,** An equilibrium is achieved when each substance is equally concentrated on both sides of the membrane and there is no further *net* movement in either direction.

Osmosis and Osmotic Pressure

None of the biological barriers is freely permeable to *all* substances. The permeability of cell membranes is so highly selective that they often appear to display the property of *semipermeability*. An ideal semipermeable membrane is one that is permeable *only* to the solvent (water), and impermeable to all solutes in a solution. Cell membranes are not ideally semipermeable by any means, but we shall temporarily consider them as such for the purposes of this discussion.

semipermeability

If a semipermeable membrane separating two different solutions allows only water molecules to move through it, then water will diffuse from a region where water molecules are *more* concentrated to a region where water molecules are *less* concentrated. *The net diffusion of water molecules between two compartments separated by a semipermeable membrane is called osmosis.*

osmosis

With respect to osmosis, it is important to understand that the number of water molecules in a solution depends on the number of solute particles in that solution. A standard volume of a 20% salt solution is considered to contain 20% salt (solute) and 80% water (solvent); a 1% salt solution would likewise contain 99% water. If these two solutions were separated by a membrane permeable only to water, the water molecules would diffuse down their concentration gradient from the more dilute salt solution (99% water) to the more concentrated salt solution (80% water). The movement of *water* in osmosis is therefore considered to take place from a dilute solution to a more concentrated solution; this statement may sound wrong, because in speaking of the strengths of solutions, we commonly refer to the solutes—not to the water—which necessarily bears the opposite relationship.

Osmosis is illustrated in Fig. 9-5. Note than an equal distribution of solute and solvent particles on both sides of the semipermeable membrane cannot be attained when only water molecules are allowed free passage. Instead, the height of the fluid in the compartment containing the more concentrated solution (less water) continues to rise as a result of the net diffusion of water into it. In theory, water molecules could continue to diffuse forever from the dilute solution into the concentrated solution, since an equal distribution, or equilibrium, of solvent and solute on both sides of the membrane could never be established. In practice, however, the net osmosis of water molecules ceases when the increasing pressure on the membrane from the weight of the rising column of fluid in the more concentrated compartment counterbalances the pressure of the water molecules diffusing in. This counterbalancing pressure is called *osmotic pressure*. The os-

osmotic pressure

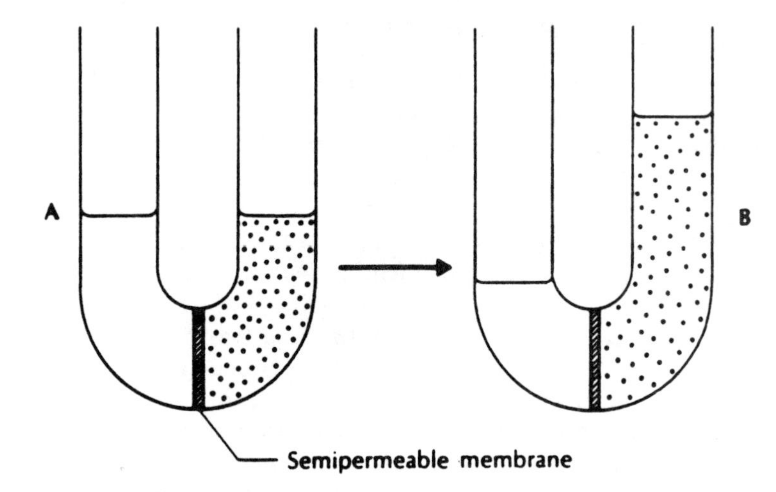

Fig. 9-5. Demonstration of osmosis. **A,** Water on the left and the more concentrated solution on the right are separated by a membrane permeable only to water molecules. **B,** Water molecules have moved down their concentration gradient from the left to the right compartment. The solution in the right compartment has become less concentrated by increasing in volume. No further net movement of water molecules can occur, because of the pressure exerted on the membrane by the column of fluid in the right compartment (osmotic pressure). Note that the number of solute particles on the right has not changed, only their concentration (number/volume).

motic pressure of a solution is a measure of its tendency to "pull" water molecules into it by osmosis. The more concentrated a solution is (that is, the more solutes and the less solvent it contains), the higher will be its potential osmotic pressure. The osmotic pressure is said to be potential because the principle only operates if the solution is separated from another solution of a different concentration by a semipermeable membrane.

Osmotic Reactions of Red Blood Cells

The fact that cell membranes are effective barriers to the free diffusion of solutes but allow water molecules to move in and out by osmosis emphasizes the importance of electrolyte balance in the body fluids. Solutions of electrolytes have the highest potential osmotic activity, because electrolytes break up into separate ions when dissolved in water, and each ion exerts its own osmotic pressure.

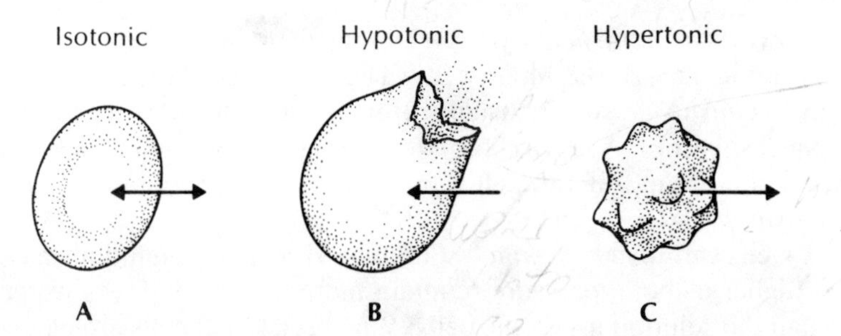

Fig. 9-6. Osmotic reactions of red blood cells. Arrows indicate the movement of water molecules. **A,** Normal cell in isotonic solution; no net movement of water into or out of the cell. **B,** Swelling of the cell in hypotonic solution due to net osmosis of water into the cell. The cell membrane has ruptured, releasing hemoglobin into the surrounding medium (hemolysis). **C,** Crenation (notched and shrunken appearance) of a red blood cell in hypertonic solution; net movement of water out of the cell.

Osmotic effects on cells are often demonstrated by suspending red blood cells in salt solutions of various concentrations (Fig. 9-6). We shall simplify this situation somewhat by regarding the red blood cell as a little sac containing a solution that is approximately 0.85% sodium chloride (NaCl). It is surrounded by a semipermeable membrane through which only water molecules can freely diffuse. Within the body, red blood cells are suspended in a fluid medium, the blood plasma, that contains the same concentrations of NaCl and water and therefore has the same osmotic pressure as the fluid inside the cells. Solutions that have the same osmotic pressure as intracellular solutions are called *isotonic,* or *isosmotic,* solutions (*iso-,* equal to). Isotonic, or *physiological, saline* is a solution that is approximately 0.85% NaCl in water. When red blood cells are placed in isotonic saline, there is no net osmosis of water molecules into or out of the cells. This is because there is no concentration gradient between the water inside the cells and the water outside the cells.

isotonic

physiological saline

Solutions in which red blood cells swell up are called *hypotonic* solutions. Hypotonic solutions have a lower osmotic pressure (contain less solutes and more water) than the solution inside the cells. When red blood cells are placed in a hypotonic solution, the water molecules outside the cell move down their concentration gradient *into* the cell. The red blood cells swell up and assume a spherical shape. If the solution is hypotonic enough (such as pure water), the cell membranes eventually rupture, releasing free hemoglobin into the surrounding medium. This phenomenon is

hypotonic

hemolysis

known as *osmotic hemolysis*. Water injected into the bloodstream would hemolyze red blood cells. Therefore, substances to be injected *intravenously* are usually prepared in isotonic saline (0.85% NaCl solution). A general term for the rupture of a cell membrane and the escape of the cell's protoplasm in hypotonic media is *plasmoptysis*.

hypertonic

Cells shrink in *hypertonic* solutions. Hypertonic solutions have a higher osmotic pressure (contain more solutes and less water) than the solution inside the cells. When red blood cells are placed in hypertonic solutions, a net osmosis of water molecules takes place from inside the cells to the outside. The red blood cells contract from loss of water, and their outlines appear wrinkled and notched. This wrinkling of red blood cells in hypertonic solutions is called *crenation*. In cells in general, it is known as *plasmolysis*.

crenation

Colloid Osmotic Pressure of Blood

Up to this point, we have considered cell membranes that allow water to diffuse freely and generally restrict the passage of solutes in true solution or colloidal solution. The osmotic pressure, or pulling pressure, of solutions on either side of a membrane is therefore dependent on *both* the crystalloids and the colloids present in solution. This situation is not quite the same when one considers the exchanges between the blood in the capillaries and the interstitial (tissue) fluid surrounding the capillaries. The capillary wall, made up of a single layer of endothelial cells, is in effect a membrane separating blood plasma (a solution containing crystalloids and colloids) from interstitial fluid (a solution containing crystalloids but only negligible amounts of colloids). The two solutions are generally in equilibrium with respect to crystalloids because the capillary wall, unlike the cell membrane, allows the free diffusion of solutes with molecular weights under about 40,000 d down their own concentration gradients. This includes all of the crystalloids and a few of the smaller colloids in blood, but it excludes the bulk of the *plasma proteins*, the higher molecular weight colloids of the blood; they are simply too large. The effective osmotic pressure of the blood in the capillaries (the difference between the blood and the tissue fluid) is therefore maintained entirely by the large colloids of the plasma and is called the *colloid osmotic pressure*. Since plasma albumin is the most abundant of the plasma proteins, the colloid osmotic pressure of the blood is mainly due to its albumin content.

plasma proteins

colloid osmotic pressure

A constant exchange of water and other substances takes place

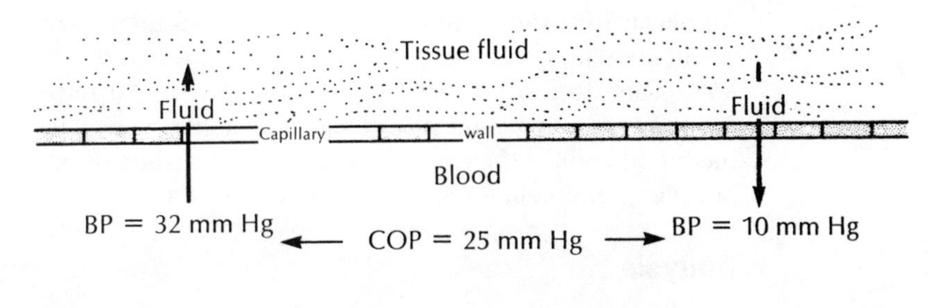

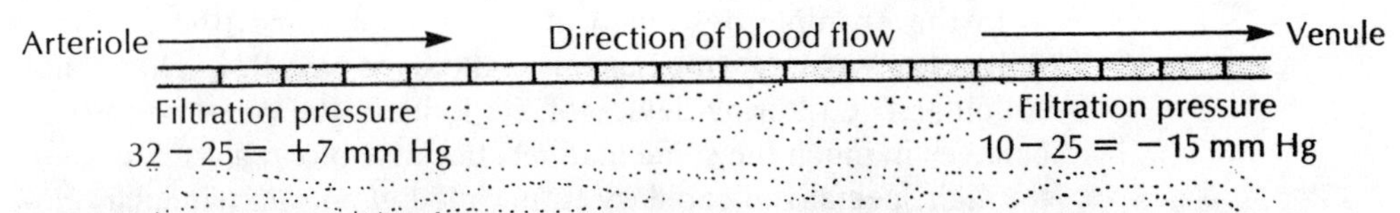

Fig. 9-7. Colloid osmotic and hydrostatic pressure relationships in a capillary. There is a net positive pushing pressure at the arteriolar end of the capillary and a net negative pulling pressure at the venular end.

between capillary blood and tissue fluid; this is the functional part of the circulatory system. The movement of fluid across the capillary wall is controlled by the interaction of the hydrostatic (fluid) pressure of blood inside the capillaries (blood pressure, BP), which tends to push fluid out of the capillaries, and the colloid osmotic pressure of the blood (COP), which tends to pull fluid into the capillaries. The difference between the two pressures, that is, BP minus COP, is the net pressure, or *filtration pressure*. The interaction of the pushing and pulling forces at the arteriolar and venular ends of a capillary is shown in Fig. 9-7. Note that although the COP is constant throughout the capillary, the BP is higher at the arteriolar end and drops significantly as the blood reaches the venular end. The resulting change in net pressure accounts for the fact that fluid *leaves* the capillary at the arteriolar end and *enters* the capillary at the venular end.

filtration pressure

If the balance between BP and COP is upset, water may accumulate in the tissue spaces. This condition is called *edema*. Two common causes of edema are an increase in the capillary BP great enough to cancel out the COP effect, and a decrease in the COP due to a decrease in its protein content. The plasma proteins, mainly albumin, may be markedly decreased in kidney disease and in some forms of malnutrition.

edema

The exchange of fluid across capillary walls is often called *bulk flow*. It is normally a rather slow process. On the other hand, individual solute ions and molecules that are within the size limit

bulk flow

for permeating the capillary wall diffuse down their concentration gradients with great rapidity *along the entire length of the capillary*. This diffusion is largely responsible for the movement of oxygen, nutrients, and physiologically active substances *out of* the blood and for the reverse movement of carbon dioxide, waste products, and cellular secretions *into* the blood.

Dialysis

In the example described above, a membrane (the capillary wall) allows mainly water and crystalloids to pass through, but not colloids. The filtering device of the kidney, the *renal corpuscle,* behaves in much the same manner, because blood circulates here through specialized capillary tufts called *glomeruli* (singular; glomerulus). The separation of crystalloids from colloids in a solution is known as *dialysis* when it is performed in the laboratory. A dialyzing membrane is one that has pores large enough to allow ions and small molecules to diffuse out but not larger colloidal particles. Cellophane is often used as a dialyzing membrane. A mixed solution of crystalloids and colloids is placed in a cellophane bag, which is then suspended in water. The crystalloids diffuse down their concentration gradients, out of the bag into the surrounding water. The crystalloids can be completely removed, leaving behind a solution containing only colloids, by using a large enough volume of water, by changing the water, or by suspending the bag in running water (Fig. 9-8).

dialysis

This principle is employed in the artificial kidney. In this case, the term *hemodialysis* is used because the blood of the patient is being circulated and dialyzed in cellophane tubes. Kidney disease causes an accumulation of toxic crystalloid waste materials in the blood. Through dialysis of the blood, these solutes can be removed without the removal of the important colloidal proteins of the blood or the blood cells. The outward diffusion of other necessary blood crystalloids, such as salt and glucose, is prevented by suspending the cellophane tubes in a solution containing the same concentrations of these solutes as found in the blood. With elimination of the concentration gradients for the essential crystalloids, there is no net loss of these substances from the blood.

hemodialysis

CARRIER-MEDIATED TRANSPORT

The movement of solvent or solute particles in diffusion, osmosis, bulk flow, and dialysis (also, the migration of charged particles in an electrical field) is *passive;* it occurs spontaneously both *in*

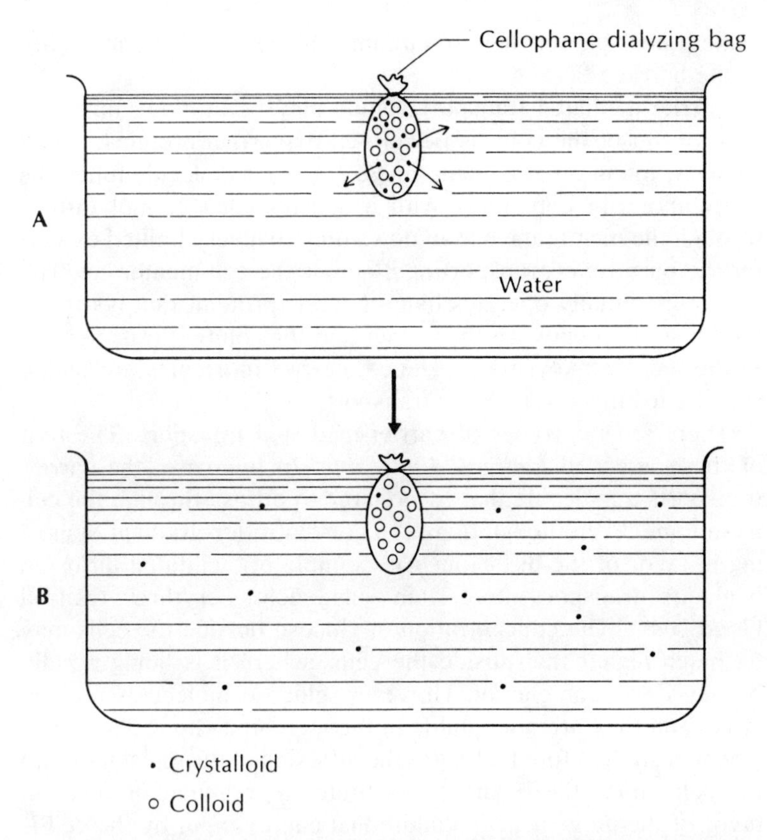

Fig. 9-8. Dialysis. **A,** A solution containing crystalloids and colloids is placed in a dialyzing bag and suspended in water. **B,** The dialyzing membrane allows the crystalloid molecules to diffuse into the water but retains the colloidal particles. The crystalloids diffuse out of the bag as long as their concentration gradient exists.

vitro and *in vivo,* according to the dictates of physical laws. Although passive movements are important factors in the membrane dynamics of living systems, the exchange of substances across cell membranes would be too limited; in fact, it would be insufficient to sustain life if all the requirements of the cell were to depend on passive transfer alone. With the exception of very small polar molecules (such as H_2O) and molecules that are soluble in the lipids of the membrane (such as the gases O_2 and CO_2 and some nonpolar organic substances), the passage of most ions and molecules is greatly restricted. Consequently, a number of carrier-mediated transport mechanisms have evolved in living cells that make it possible for them to circumvent the limited permeability of the barrier.

carriers

Carrier-mediated transport, which may be in *either* direction (into or out of the cell), is performed by certain proteins, called *carriers,* found on cell membranes. A carrier molecule functions by temporarily combining with a species (that cannot diffuse through the membrane or can do so only to a very limited extent) for the purpose of transporting it across the cell membrane. The carrier essentially operates like a binding protein, that is to say, the interaction between the carrier and the solute it transports is in the lock-and-key mode. Hence, carrier molecules are highly specific for the species they transport.

facilitated diffusion

There are two types of carrier-mediated transport. The first of these is called *facilitated diffusion.* In this case, the carrier transports a molecule that is too large to diffuse through the cell membrane, even though there is a concentration gradient operating in favor of the diffusion. An example of facilitated diffusion is glucose transport into certain cells, such as erythrocytes (red blood cells). The concentration of glucose outside the cells may be much higher than inside the cells, where it is being rapidly broken down for energy. However, glucose molecules are too large, and they are not soluble in the membrane lipids. A carrier mechanism therefore facilitates the diffusion of the molecules into the cell. Since the gradient constitutes a spontaneous force in favor of the movement, no additional energy input by the cell is required to transport the substance.

active transport

The second type of carrier-mediated transport is known as *active transport*. Here, substances are moved into or out of the cell by specific carriers *against* concentration gradients, that is, from a region of low concentration to a region of high concentration. This process is *active,* not passive, because it is in an energetically unfavorable 'uphill' direction, and it requires an input of energy. Active transport systems are thus distinguished by their ability to operate against gradients, and by their dependence on a source of energy.

uniport
cotransport

Observed differences in the modes of operation of various carrier-mediated transport systems provide another basis for classifying them (Fig. 9-9). Many carrier systems are involved in the transport of only one substance; these are called *uniports*. In other systems, collectively termed *cotransports,* the carrier cannot function unless two different substances are more or less simultaneously transported. When the coupled transport of both substances takes place in the same direction across the membrane, the system is called a *symport*. Coupled transport of the two sub-

symport

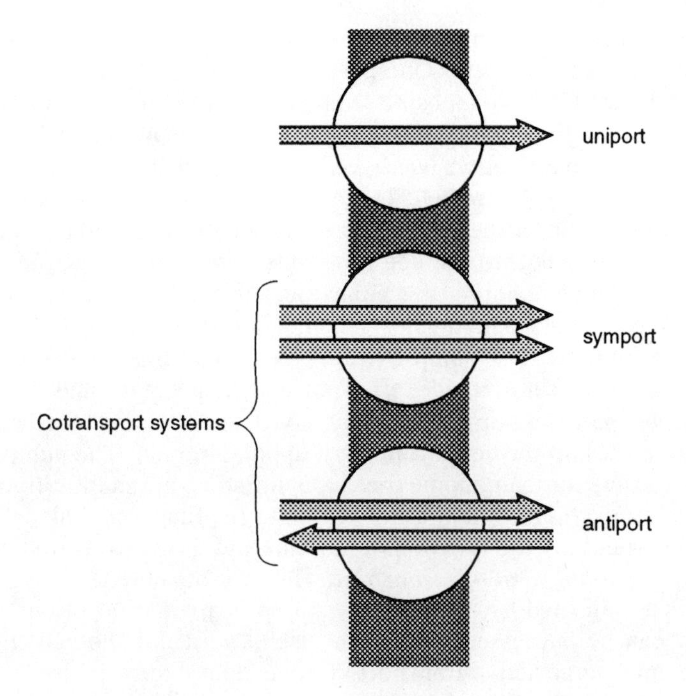

Fig. 9-9. Three types of carrier-mediated transmembrane transport. Uniport, and cotransport, the coupled transport of two substances in the same direction (symport), and in opposite directions (antiport).

stances in opposite directions across the membrane is called an *antiport*.

antiport

Ion Pumps

The carrier proteins that mediate the active transport of ions are commonly called 'pumps'. One much-studied pump, the sodium-potassium (Na^+-K^+) pump (discussed below), controls the intracellular concentrations of the two predominant cations of the body fluids, Na^+ and K^+. Calcium pumps maintain low levels of intracellular Ca^{2+}, as compared to those outside the cell, and are also responsible for sequestering these physiologically active ions within intracellular subcompartments. Other pumps are used for concentrating ions in exocrine secretions, such as H^+ in gastric juice, and bicarbonate ions in pancreatic juice.

ion pumps

The Na^+-K^+ pump. A notable example of active ion transport is the widely distributed antiport system that mediates the coupled uphill transport of Na^+ and K^+ in opposite directions across cell

Na^+-K^+ pump

membranes. The pump is a high-molecular weight multisubunit transmembrane protein. One part of the complex is an enzyme, Na^+-K^+ ATPase (adenosine triphosphatase) which hydrolyzes ATP, making energy available for the ion transport. Recall that the predominant cation *inside* cells is K^+ and the predominant cation *outside* cells is Na^+. There is thus a constant tendency for K^+ to leak out of the cell (down its concentration gradient) and for Na^+ to leak into the cell (down its concentration gradient). Indeed, if it were not for this pump mechanism, the concentration of the two ions would quickly reach equilibrium on both sides of the membrane. The pump extrudes Na^+ ions that have leaked into the cell from outside, transporting them uphill, and on its return trip, it picks up K^+ ions that have leaked out and transports them back into the cell, again in an uphill direction. The pump is responsible for maintaining the characteristic Na^+ and K^+ transmembrane gradients that are essential to the function of all cells, and particularly the function of muscles and the nervous system.

Na^+-linked active transport

Sodium-linked active transport. The transmembrane Na^+ gradient established by the Na^+-K^+ pump is a reservoir of energy that can be harnessed to perform useful work for the cell. For example, some active transport systems do not directly use ATP as a source of energy, but are driven instead by the energy produced by the flow of Na^+ ions down their gradient. Some well-known examples of this are the Na^+-linked symports used by epithelial cells lining the small intestine. The latter cells are specialized for uphill transport, that is, their primary function is to absorb and accumulate end products of digestion, such as glucose and amino acids, from the lumen of the intestine. *(Note, in other types of cells, glucose is transported by a facilitated diffusion mechanism; see above.)* In these symports, Na^+ ions and the glucose or amino acid molecules to be transported simultaneously bind to the carrier, and are cotransported into the cell. The 'fall' of the Na^+ ions *down* their steep transmembrane gradient supplies the energy to drive the *uphill* transport of the other substance. The Na^+-K^+ pump is obviously a key factor here since it pumps out the Na^+ ions that have entered the cell, and thus reestablishes the gradient necessary for the continued operation of the symport. In this context, Na^+-linked active transport systems *indirectly* consume energy in the form of ATP. Sodium-linked symports are sometimes termed *secondary active transport*.

ENDOCYTOSIS AND EXOCYTOSIS

Macromolecules and particulate material are normally too bulky to get through or across an intact cell membrane. Nevertheless, there is a constant traffic of such materials into and out of cells. The cellular mechanisms used for this purpose are termed *endocytosis* and *exocytosis*.

endocytosis
exocytosis

Endocytosis (*endo*-inside; *kytos*-cell) is the process by which large bulky materials enter cells, and exocytosis (*exo*-outside) is the process by which they are extruded from cells. Both processes involve physically breaking through the cell membrane and, like active transport, both require an expenditure of energy by the cell. In endocytosis, the material taken into the cell is enclosed in a patch of cell membrane which then pinches or buds off the main body of the membrane, and enters the cytoplasm as a membrane-enclosed vesicle. When the budding-off process is completed, the cell membrane reseals itself and closes up the gap. This wrapping of the endocytosed material in bits of membrane prevents it from coming in direct contact with the intracellular fluid, or cytosol. Exocytosis similarly involves vesicle formation but, in this case, the vesicles bud off from the membranes of various intracellular organelles, such as the Golgi apparatus and the endoplasmic reticulum, and subsequently fuse with the cell membrane.

Endocytosis

As shown in Fig. 9-10, three mechanisms have evolved for the internalization of large-sized materials: these are *phagocytosis* (cell 'eating'), *pinocytosis* (cell 'drinking'), and *receptor-mediated endocytosis*.

phagocytosis
pinocytosis
receptor-mediated endocytosis

In phagocytosis, the cell ingests a solid particle or a large molecular assembly. When a cell comes into contact with material of this kind, the cell membrane expands around it and closes over it. The patch of membrane with the enclosed material then buds off the cell membrane and enters the cytoplasm as a phagocytotic vesicle, or *phagosome*. Most phagosomes are fairly large vesicles that may measure several micrometers in diameter. Protozoa, such as amoebae, commonly use this process as a form of feeding. However, many cells in humans and other mammals also exhibit phagocytic activity. Among the most active phagocytic cells in the body are cells of the immune system, such as neutrophils (a species of white blood cells, or leukocytes) and macrophages. These cells use phagocytosis to engulf and destroy invading bacteria and viruses, and also to ingest dead, senile, or damaged cells, and bits of cellular debris. The ingested material is usually destroyed by the hydrolytic enzymes in cell organelles called lysosomes.

Pinocytosis is the process by which the cell internalizes a tiny

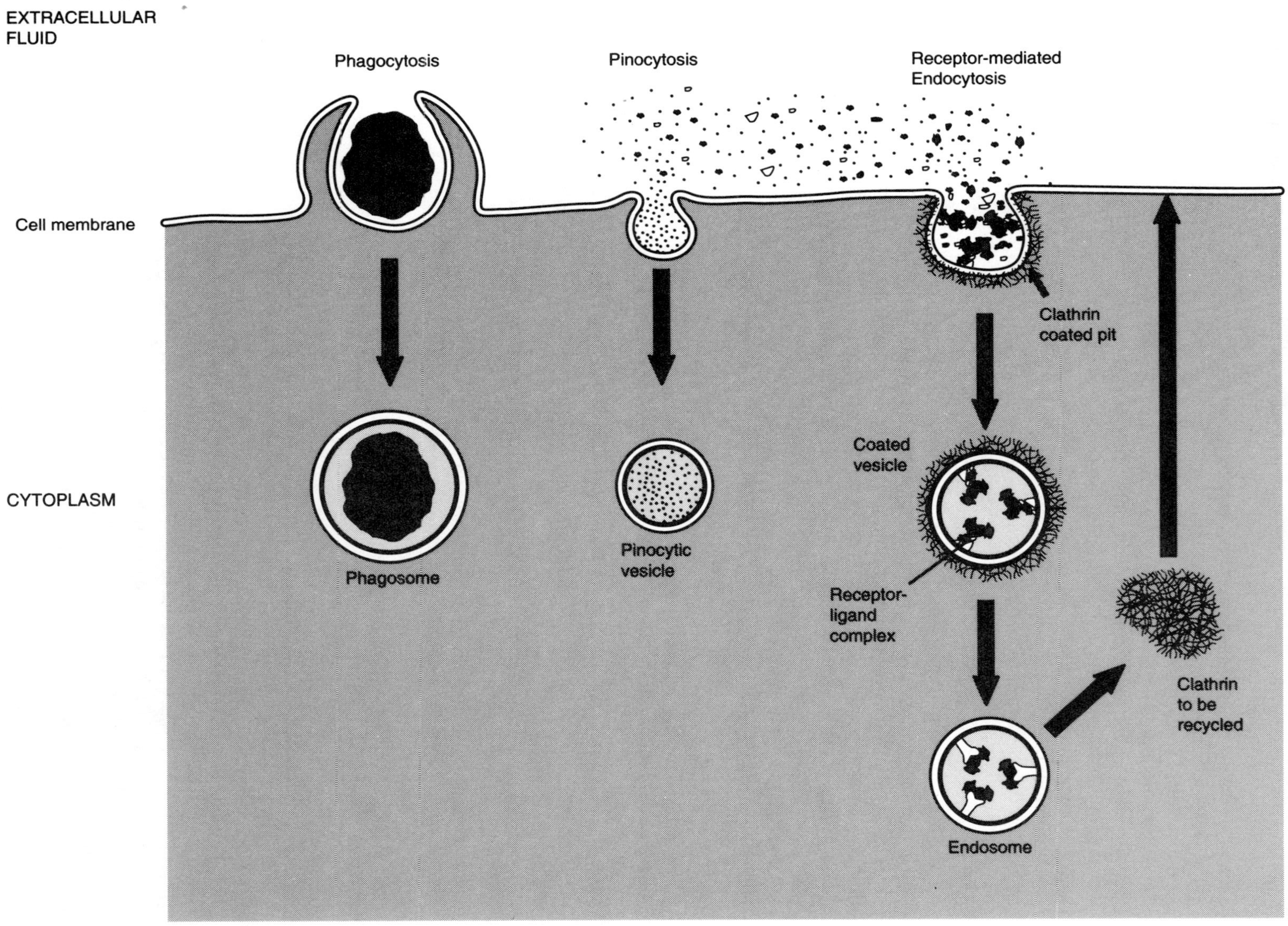

EXTRACELLULAR
FLUID
Phagocytosis
Pinocytosis
Receptor-mediated
Endocytosis
Cell membrane
Clathrin
coated pit
Coated
vesicle
CYTOPLASM
Phagosome
Pinocytic
vesicle
Receptor-
ligand
complex
Clathrin
to be
recycled
Endosome

droplet of extracellular fluid. The mechanism here is similar to phagocytosis, that is, the droplet is enclosed in a piece of invaginated membrane, and internalized in the form of a vesicle. Pinocytosis is a nonselective process because any solutes that happen to be present in the fluid are incorporated into the pinocytic vesicle. The uptake of dissolved materials by this route thus depends solely on how concentrated they are in the extracellular fluid. Pinocytic vesicles tend to be very small, generally below the resolution of light microscopy. The process is common in all mammalian cells.

In contrast to fluid-phase pinocytosis, the process of receptor-mediated endocytosis is highly selective because it involves the internalization of specific membrane receptor-ligand complexes. Membrane receptors are typical binding proteins, many of them transmembrane proteins, which are positioned in the cell membrane so that their binding sites face the extracellular fluid. Since their binding sites are stereospecific, receptor proteins interact only with those ligand molecules that 'fit' into them, in other words, a lock-and-key interaction. They can thus ignore the multitude of other substances floating around in the extracellular fluid, and selectively bind one particular species of ligand, even though it may be present in very low concentrations. The receptors that are apparently earmarked for internalization are those clustered in special regions of the cell membrane, known as *coated pits*. The bristle-like coating around these pits is made up of a high-molecular weight fibrous protein called *clathrin*. Studies of coated pits have revealed that they have an average area of roughly 0.1 μm^2, and may contain about 1000 membrane receptor molecules of different types. Coated pits, with their content of receptor-ligand complexes, bud off the cell membrane to form *coated vesicles*. As the latter move deeper into the cell, they shed their clathrin coats and become smooth-surfaced vesicles, called *endosomes*. These vesicles undergo various types of processing within the cell, but clathrin and the receptors are evidently recycled back

coated pits

clathrin

coated vesicles

endosomes

←

Fig. 9-10. Schematic representation of the three modes of endocytosis. In phagocytosis on the left, a large particle is enclosed in cell membrane and internalized as a membrane-bound vesicle called a phagosome. Pinocytosis in the center illustrates the internalization of a fluid droplet with the various small solutes it contains. In receptor-mediated endocytosis on the right, specific ligands in the extracellular fluid are bound to membrane receptors in a coated pit. The receptor-ligand complexes are then internalized in clathrin-coated vesicles. The clathrin then comes off the vesicle, leaving a smooth-surfaced vesicle called an endosome.

to the cell membrane to be used over again. There is now a growing list of materials that are taken into cells via receptor-mediated endocytosis. These include hormones, LDL, growth factors, antibodies, transferrin (an iron-transporting plasma protein), and bacterial toxins. Certain viruses also apparently use this mechanism to invade cells.

Exocytosis

In this process, a vesicle containing one or another substance fuses with the cell membrane and discharges its contents to the exterior (Fig. 9-11). Exocytosis principally occurs in cells that are specialized for endocrine, exocrine, and other types of secretion. The secretory products of such cells are usually synthesized in the rough endoplasmic reticulum (rER), and then transferred to the Golgi apparatus for further processing. The finished products are eventually packaged in membrane-bound *secretory vesicles* which are often seen budding off the membrane of the Golgi apparatus. Characteristically, cells specialized for secretion have a well-developed endoplasmic reticulum and Golgi apparatus, and numerous secretory vesicles in their cytoplasm. Once they have been formed, secretory vesicles either fuse with the cell membrane and discharge their contents from the cell, or remain in the cytoplasm until an appropriate signal or stimulus triggers the exocytosis.

secretory vesicles

It should be noted here that budding and fusion events on the cell membrane, and on subcellular membranes, are involved in all types of endocytosis and exocytosis. Membrane turnover, and exchanges of membrane material among the various membranes of the cell, are constant features of cell function. In all these events, however, the asymmetry and the characteristic architecture of each membrane are preserved.

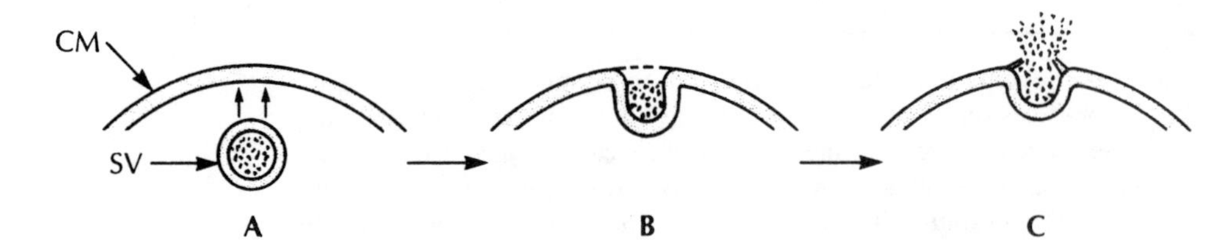

Fig. 9-11. Exocytosis (secretion). **A,** The membrane-enclosed secretory vesicle *(SV)* moves toward the cell membrane *(CM)*. **B,** The two membranes fuse, **C,** The secretory product is discharged from the cell.

10

ENERGY RELATIONS IN CELLS

It is common knowledge that man must eat and breathe in order to live. In fact, eating and breathing are interrelated: the uptake of atmospheric oxygen is required for the liberation of life-sustaining energy from food. Exhaling the end product, carbon dioxide, is the final expression of that reaction. The liberation of energy from food by the process of *cellular* (internal) *respiration* is a chemical function carried out by the cells of all living organisms.

cell respiration

In larger multicellular animals, the direct exchange of nutrients and respiratory gases between the external environment and the body cells is not possible. *Breathing* (external respiration) has evolved as a mechanical means of getting O_2 into the body and CO_2 out of the body. Digestive and circulatory systems are accessories in delivering the necessary substrates to the cells and removing the end products of respiratory reactions.

breathing

METABOLISM

Cell respiration is an integral part of cell metabolism. The term *metabolism* is defined as the sum total of all the chemical and physical (energy) changes that occur in the cells of the body. The metabolic activity of cells is divided into: (1) *catabolism,* the breaking of the chemical bonds in large, more complex molecules (such as food molecules) to form smaller, more simple molecules and (2) *anabolism,* the biosynthesis of larger molecules by chemically bonding together smaller molecules.

metabolism

catabolism

anabolism

In general, catabolic processes *liberate* energy *(exergonic reactions),* because the energy content of a large molecule is greater than that of its smaller breakdown products. Anabolic processes generally *consume* energy *(endergonic reactions).* Muscle con-

exergonic

endergonic

traction, nerve conduction, maintenance of body temperature, and active transport also use up energy. These aspects of metabolism imply the existence of a dynamic energy exchange mechanism; that is, energy released by the catabolism of food molecules is utilized by energy-consuming processes.

CHEMICAL BOND ENERGY

Energy is defined as the capacity to do work. It may be expressed in the form of heat and light or mechanical, electrical, or chemical energy. Although, according to the law of conservation of energy, it can neither be created nor destroyed during chemical or physical reactions, energy can be transformed from one kind to another under the proper conditions.

The body depends on an adequate supply of energy for the maintenance of life processes. The question is: What form of energy can it use for this purpose? Animals do not possess the necessary mechanisms to make biological use of heat, light, electrical, or mechanical energy. The only form of energy available to them is the potential energy stored in the chemical bonds of food molecules.

photosynthesis

The ultimate source of chemical bond energy is the light emitted by the sun. But, of all life forms on earth, only plants are able to convert solar energy *directly* to chemical bond energy; they do so by the process of *photosynthesis*. Man thus consumes the energy of the sun *indirectly* by virtue of his position at the top of a food web; that is, he eats both the plants and other animals that feed on plants.

Photosynthesis is the sum of a complicated series of chemical reactions that can be summarized as follows:

$$6CO_2 + 6H_2O \xrightarrow[\text{chlorophyll}]{\text{Light and}} 6O_2 + C_6H_{12}O_6$$

chlorophyll
excited electrons

In brief, the photons of light rays transfer their energy directly to electrons in the *chlorophyll* molecule (the green pigment of plants). The plant cells then remove the energy from the *excited electrons* and use it to synthesize glucose ($C_6H_{12}O_6$). In most photosynthetic plants, this is the starting point for the manufacture of the other organic molecules of vegetable origin. The raw materials for photosynthesis are the gas carbon dioxide (CO_2), which is present in small amounts in the atmosphere, and water (H_2O). It should be noted that oxygen given off in photosynthesis is the source of the oxygen in the earth's atmosphere.

The light energy invested in the synthesis of glucose from carbon dioxide and water is locked into the bonds that hold the carbon, hydrogen, and oxygen atoms together within the glucose molecule. That energy can be released only when the bonds are broken. Living cells extract the energy by chemically breaking down food molecules to carbon dioxide and water (essentially reversing the work of photosynthesis). The energy changes that accompany these biochemical reactions are measured in terms of *free energy,* that is, energy that can do useful work (at constant temperature and pressure).

free energy

ENERGY UNITS

Although there are various units for measuring the different forms of energy, the main unit used for energy exchanges of metabolism is a heat energy unit, the *calorie*. A calorie is the quantity of heat energy required to raise the temperature of 1 gram of water by 1 degree Celsius. The energy content of foods is often expressed by the large Calorie (written with a capital C). The large Calorie is actually a kilocalorie (kcal), equal to 1,000 small calories.

calorie

The free energy changes that occur during metabolic reactions in cells are quantitatively expressed as *kcal per mole* of the substance(s) undergoing the chemical change(s). If the free energy change in a substance is negative (−kcal/mole), the reaction liberates energy and is exergonic (''downhill''); if it is positive (+kcal/mole), the reaction is endergonic (''uphill'') and requires an investment of free energy before it will take place.

HIGH-ENERGY COMPOUNDS OF THE CELL

The transfer of free energy from catabolic energy-producing reactions to energy-consuming reactions is mediated in living cells by chemical compounds that are able to *store energy in high-energy phosphate bonds*. The key high-energy compound of this group is *adenosine triphosphate (ATP),* a nucleotide formed by the nitrogen base adenine, the sugar ribose, and three phosphate groups bonded in sequence (nucleotides are phosphoric acid esters of nucleosides; for further details of these important biological compounds, see Chapter 11). The compound ATP can be represented as

high-energy bonds

ATP

$$A—P{\sim}P{\sim}P$$

The notation ~P conventionally indicates the high-energy bond. On hydrolysis, ATP loses its terminal phosphate group and is converted to *adenosine diphosphate (ADP)*, with the release of about 7,300 calories (7.3 kcal) of free energy per mole*:

ADP

$$A—P{\sim}P{\sim}P + H_2O \rightarrow A—P{\sim}P + P_i + \text{energy}$$

P_i stands for inorganic phosphate groups that are present in abundance in all cells. Approximately the same amount of energy can be released by the hydrolysis of ADP to AMP (adenosine monophosphate), but this reaction occurs much less frequently in cells.

The conversion of ATP to ADP is an exergonic reaction in which free energy is released. ATP is resynthesized from ADP by the process of *phosphorylation* (addition of a phosphate group). This reverse reaction is endergonic and requires an input of energy to reconstitute the terminal high-energy phosphate bond.

phosphorylation

$$\textit{Exergonic:}\ ATP \rightarrow ADP + P_i + \text{energy}$$

$$\textit{Endergonic:}\ ADP + P_i + \text{energy} \rightarrow ATP$$

The advantages of such an economical and efficient energy exchange system are evident. ATP is the vital link between catabolism, on the one hand, and energy-consuming cell activities, on the other (Fig. 10-1). The cycle can be summarized as follows:

1. Chemical energy released by exergonic reactions is used for the phosphorylation of ADP to energy-rich ATP.
2. The energy stored in ATP is used by the cell as a source of power for the work of the cell: active transport, biosynthesis, muscle contraction, and so on. The ATP used in this manner gives up its terminal energy-rich phosphate bond and is converted to ADP.
3. Repeat 1; ADP formed in 2 is phosphorylated; and so on.

From the foregoing, it is evident that the phosphorylation of ADP is a pivotal event in metabolism. As we shall see, cells employ two mechanisms to phosphorylate ADP. In the first of these, termed *substrate-level phosphorylation*, a high-energy phosphate bond is directly generated from an energy-rich substrate in the course of a metabolic reaction. The second mechanism, *oxidative phosphorylation*, is the principal source of ATP (and energy) for

substrate-level phosphorylation

oxidative phosphorylation

* This measurement has been arbitrarily estimated on the basis of certain standard conditions and may be lower than the actual amount of energy released in the intact cell.

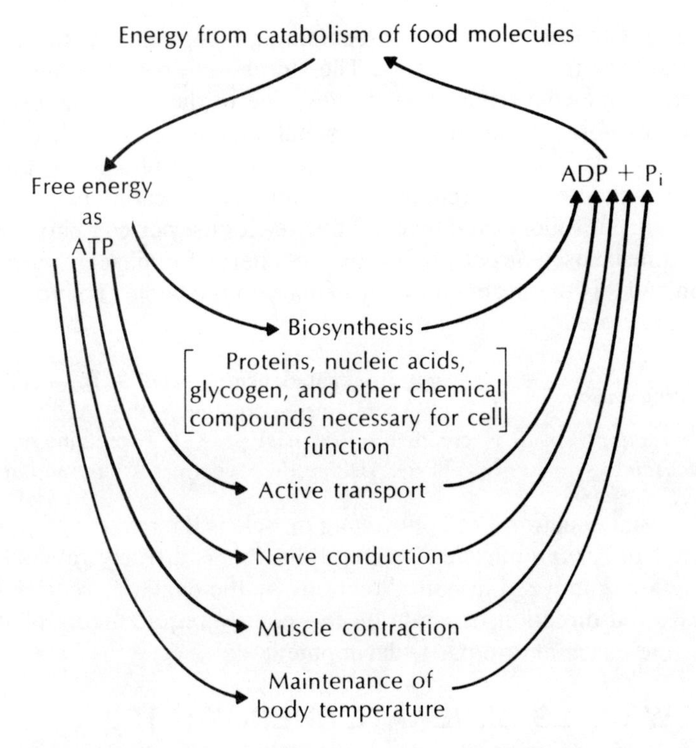

Fig. 10-1. The oxidation of food substrates in the cell provides the energy for the phosphorylation of ADP to ATP. The energy in ATP is then used for carrying out energy-consuming activities.

human cells, and indeed for all oxygen-consuming cells. In this case, the formation of ATP is coupled to the oxidation of hydrogen carriers on the electron transport chain in mitochondria (see below). The final electron acceptor in this mechanism is O_2, and the end product here is H_2O.

Although ATP is the major high-energy compound of all living systems, other high-energy compounds are used in special circumstances. Related nucleoside triphosphates, such as *guanosine triphosphate (GTP)* and *uridine triphosphate (UTP)*, participate in certain metabolic reactions. For example, GTP is generated by substrate-level phosphorylation in the citric acid cycle. Phosphate groups from the nucleotides can be interchanged in the presence of the appropriate enzymes, as follows:

GTP
UTP

$$\text{GTP} + \text{ADP} \rightleftharpoons \text{ATP} + \text{GDP}$$

Another energy-rich phosphate compound, found mainly in muscle tissues, is *phosphocreatine* (also known as creatine phos-

phosphocreatine

phate). Creatine is a nitrogen-containing compound synthesized in the liver from amino acids. The storage of energy in muscles varies somewhat from other tissues, due to the fact that resting muscles normally maintain only small supplies of ATP. Energy for muscular work is stored in the high-energy phosphate bond of phosphocreatine (creatine ~ P). However, creatine cannot be *directly* phosphorylated by catabolic reactions, nor can phosphocreatine be used *directly* as a source of energy for muscle contraction. ATP is the necessary intermediary in the cycle. The following illustrates the events in muscles:

Resting muscle	1. energy from catabolism + ADP + P_i → ATP
	2. ATP + creatine → creatine ~ P + ADP
Contracting muscle	1. creatine ~ P + ADP → ATP + creatine
	2. ATP → ADP + P_i + energy for contraction

Note that equation 1 of *contracting* muscle is the reverse of equation 2 of *resting* muscle. In operation, this is a freely reversible reaction, catalyzed in both directions by the enzyme *creatine kinase;* the direction depends on the energy requirements of the muscle tissue at any particular moment.

HOW CELLS GENERATE ENERGY (ATP)

oxidation

An important process in the extraction of energy from chemical bonds is oxidation. Technically, *oxidation* is the loss of electrons from an atom or a compound. Chemical reactions generally do not involve free electrons; therefore, every electron lost by oxidation in one compound must be gained by another compound. The gain of electrons is called *reduction.* Oxidation and reduction thus occur simultaneously and may be considered as a competition between two substances for electrons; when one substance becomes oxidized, the other is reduced, and vice versa.

reduction

In biological oxidation-reduction reactions, electrons and protons are often lost and gained together—as hydrogen atoms (H). (A proton and an electron make an H atom.) Thus, compounds are said to be oxidized when they *lose* 1 or more atoms of hydrogen and reduced when they *gain* 1 or more atoms of hydrogen. In fact, hydrogen atoms are usually lost and gained (exchanged, actually) *in pairs.*

Hydrogen (Electron) Carriers

The overall reaction by which cells extract free energy from the oxidation of food molecules can be summarized as follows, with glucose as the representative food molecule:

$$C_6H_{12}O_6 + 6O_2 \rightarrow 6CO_2 + 6H_2O$$

Hydrogen atoms are removed (glucose is *oxidized*), the bonds of the glucose molecule are broken, and carbon dioxide and water are formed as the end products. The carbon dioxide represents the maximally oxidized C atoms of the original glucose molecule; the water results from molecular oxygen being reduced by the hydrogen atoms from the glucose.

oxidizing agent

The molecular oxygen supplied by external respiration (breathing) of animals is the ultimate *oxidizing agent* of glucose; but oxygen does not react directly with glucose, as it does when glucose is burned in a flame. The same amount of energy is liberated in cell respiration, but not all at once. Rather, it is siphoned off a little at a time by many individual chemical reactions. Total rapid release of the energy would destroy the cell. A useful analogy is the means whereby a person could descend from the top story of a building to street level. He could accomplish his goal either by jumping out of a window or by using a stairway. The first method would achieve his descent rapidly and effectively, but at the price of his life; the second alternative, although a slow, step-by-step process, would deliver him to ground level in an intact and healthy condition. The many steps of cell oxidation form a stairway by which the high bond-energy levels of the glucose molecule descend to the low bond-energy levels of carbon dioxide and water.

NAD^+

FAD

The principal carriers that accept hydrogen and its electrons directly from glucose and the chemical intermediates produced during its stepwise oxidation are the coenzymes *nicotinamide adenine dinucleotide (NAD^+)*, and, to a lesser extent, *flavin adenine dinucleotide (FAD)*. These coenzymes, which are composed of vitamins of the B complex and nucleotides, operate in conjunction with a group of enzymes classified as *dehydrogenases* (refer to Table 8-1, Chapter 8, and discussion of nucleotides in Chapter 11). The chief hydrogen acceptor is NAD^+, which actually carries 1 H^+ and 2 electrons in the following manner (leaving 1 proton loose in the cellular fluids):

$$NAD^+ + 2H \rightleftharpoons NADH + H^+$$

However, to simplify the discussion, the reduced form of this hydrogen carrier will at various times be indicated as $NADH_2$. FAD does in fact carry 2 hydrogen atoms; its reduced form is $FADH_2$.

After being reduced by hydrogen atoms of the substrates, the carriers in turn reduce a series of electron-transferring substances (the enzymes and coenzymes of the electron transport system), which eventually reduce molecular oxygen to form H_2O.

Pathway of Oxidation

Carbohydrates, fats, and proteins are the essential components of man's diet. All these food molecules serve as substrates in cell oxidation. However, since the oxidation of carbohydrate, specifically in the form of glucose, is our most important source of energy, and all food molecules are ultimately oxidized by the same pathway, this discussion will be based on the breakdown of the glucose molecule.

Biological oxidation in mammalian cells occurs in three stages:

glycolysis

STAGE I: *Glycolysis (Embden-Meyerhof pathway)*
Definition: The series of reactions by which 1 molecule of glucose is converted to 2 molecules of pyruvic acid (pyruvate).

anaerobic

Characteristics: Does not require oxygen (is *anaerobic*); takes place in the cytosol (cytoplasmic fluid of the cell).

citric acid cycle

STAGE II: *(a) Conversion of pyruvic acid to acetyl coenzyme A and (b) citric acid cycle (Krebs cycle; tricarboxylic acid cycle)*
Definition: The chemical machinery for the complete oxidation of acetyl-CoA (activated acetic acid), the breakdown product of pyruvic acid.
Characteristics: Requires the presence of oxygen (is *aerobic*); takes place in the mitochondria of the cell.

electron transport chain
oxidative phosphorylation

STAGE III: *Electron transport chain and oxidative phosphorylation*
Definitions: The *electron transport chain* is the final pathway for all electrons removed from substrate molecules during oxidation. *Oxidative phosphorylation* is the process in which the oxidation of the hydrogen electrons removed from substrates is coupled to the phosphorylation of ADP (to ATP).
Characteristics: Requires the presence of oxygen (aerobic); takes place in the mitochondria of the cell.

Characteristic features of biological oxidation reactions are:

1. They involve a stepwise, gradual sequence of complex chemical reactions.
2. Every step is catalyzed by enzymes.
3. The overall process is exergonic (''downhill'') even though some steps are endergonic, notably the phosphorylation of ADP to ATP.

4. Six molecules of CO_2 and 12 molecules of H_2O are the final products of each completely oxidized molecule of glucose ($C_6H_{12}O_6$). Since 6 molecules of H_2O are invested in the oxidation, the *net* yield of H_2O is actually 6 molecules. It should be noted that the *molecular oxygen* breathed in during the process of external respiration acts *only* as an acceptor of *hydrogen atoms* (forming the final product, H_2O)—the oxygen atoms in CO_2 are derived from water, which plays the role of a reactant in several steps of the oxidative pathway.
5. By convention, the complete oxidation of 1 mole of glucose (or 1 molecule, since the *ratio* will be the same) is considered to yield enough energy to make a total of 36 moles (or molecules) of ATP, that is, a ratio of 1:36. It should be understood, however, that this is a variable ratio dependent on chemical conditions inside the living cell. Thus, during any particular phase of cell activity, the *actual* yield of ATP could vary from about 25 to 45 moles per mole of glucose oxidized. Note that even the lowest estimated yield (25 ATP) is still highly favorable to the cell.

Glycolysis

In glycosis (meaning dissolution of sugar), 1 molecule of glucose is split into 2 molecules and partially oxidized to pyruvic acid by a series of about nine reactions. ATP is formed directly by substrate-level phosphorylation in two of the reactions. At this starting point, it is important to keep in mind the fact that the large number of organic acids produced at various stages of the oxidative pathways, for example, pyruvic acid, lactic acid, phosphoglyceric acid, citric acid, and so on, exist in the cell in the form of their anions. Thus, the more accurate terms for them are pyruvate, lactate, phosphoglycerate, citrate, and so on. For the purpose of clarity, we shall use the full names of the acids, with occasional reference to relevant anions where convenient.

The important steps of glycolysis are shown in Fig. 10-2 and proceed in the following order:

1. The stable glucose molecule is first activated (primed) by phosphorylation, the addition of a phosphate group. The cell uses one ATP for this purpose:

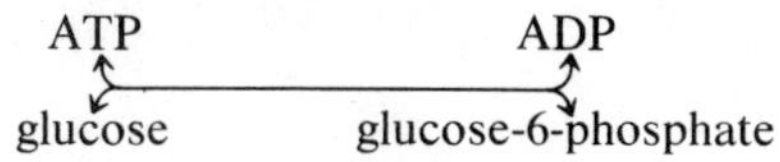

ATP → ADP

glucose → glucose-6-phosphate

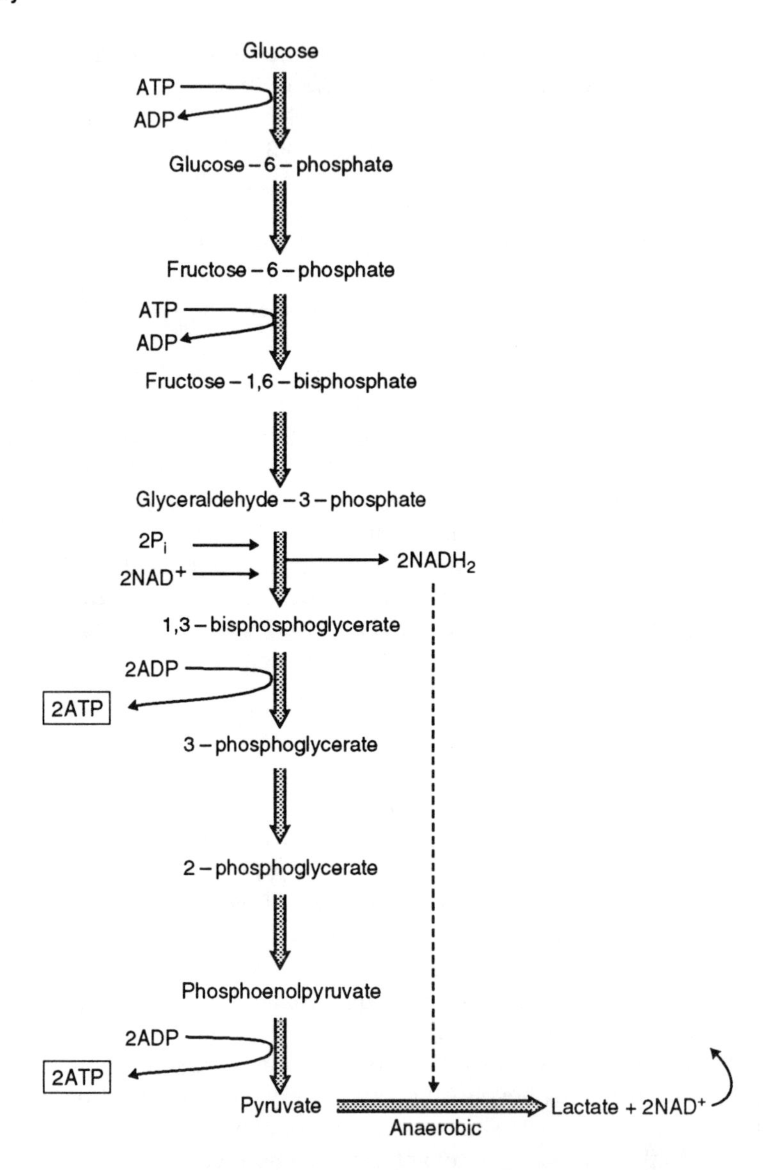

Fig. 10-2. The reactions of glycolysis, the conversion of glucose to pyruvate (pyruvic acid). Note, glucose and fructose are 6-carbon compounds, and all the substrates formed after fructose 1,6-bisphosphate are 3-carbon compounds. Hence, *two* molecules of all substrates from glyceraldehyde 3-phosphate to lactate (lactic acid) are formed for each glucose molecule fed into the glycolytic pathway. Note also, the generation of ATP by substrate-level phosphorylation at two reactions in the pathway.

2. Glucose-6-phosphate is rearranged to its more symmetrical isomer, fructose-6-phosphate; then a second phosphate group is added by investing a second ATP:

ATP → ADP

fructose-6-phosphate → fructose-1.6-bisphosphate

3. Fructose-1,6-bisphosphate is split into 2 molecules of glyceraldehyde 3-phosphate, a 3-carbon compound. The cell, which up to this point has only *invested* energy, now proceeds to harvest some energy from the substrates on hand.
4. In the next step, two molecules of the hydrogen carrier, $NADH_2$ (actually NADH), are produced when 2 molecules of glyceraldehyde 3-phosphate are oxidized to (2) 1,3 bisphosphoglycerate (1,3 BPG). The latter intermediate has a high-energy phosphate group. It reacts with ADP to form (2) ATP, generated here by substrate-level phosphorylation. 1,3 BPG is then hydrolyzed to 3-phosphoglycerate.
5. The reaction that follows is a conversion of 3-phosphoglycerate to 2-phosphoglycerate by changing the position of the phosphate group in the molecule.
6. A molecule of water is then removed from 2-phosphoglycerate, forming phosphoenolpyruvate (PEP) which has a high-energy phosphate group. Its phosphate is transferred to ADP to produce ATP, once again by substrate-level phosphorylation, and pyruvate (pyruvic acid) is formed in the process.

In summary, the process of glycolysis can be represented as:

$$\text{glucose} + 2\text{NAD}^+ + 2\text{ADP} + 2\text{P}_i \rightarrow 2\,\text{pyruvate} + 2\text{NADH}_2 + 2\text{ATP}$$

Note from the above discussion and Fig. 10-2 that the 2 ATP molecules expended in priming the glycolysis pathway are re-formed later in the sequence, along with 2 additional molecules of ATP, representing a net gain of 2 ATP. Even more of the energy of the original glucose molecule is conserved in the 2 $NADH_2$ molecules formed during glycolysis. If it is available, molecular oxygen will eventually reoxidize these reduced coenzymes, producing 4 more ATP molecules in the process (see electron transport discussion, p. 194.) However, if there is no oxygen present,

lactate

the hydrogen atoms from $NADH_2$ are transferred to pyruvic acid, forming *lactic acid (lactate).*

Lactic Acid Formation

recycled NAD^+

NAD^+ is present in very minute amounts in the cell. If after being reduced, it could not be reoxidized and thus made available for re-use, glycolysis would soon be stalled. The reduction of pyruvic acid to lactic acid provides a mechanism for *recycling NAD^+* in the absence of oxygen (Figs. 10-2 and 10-3).

Glycolysis does not provide very much ATP from a molecule of glucose (only 2 molecules), but, if necessary, the pathway can operate very rapidly and serve temporarily as the sole source of energy. For example, in actively contracting muscle, sufficient oxygen is not always available. Such deficiency may occur because the physical diffusion of oxygen to the reaction site is a slow process compared with the rate of chemical reaction and muscle contraction.

oxygen debt

The formation of lactic acid therefore frees sufficient NAD^+ to enable glycolysis to proceed in the presence of little or no oxygen. Lactic acid that accumulates in muscles can be reconverted to pyruvic acid, but only after the oxygen stress has passed, because oxygen is required for this reconversion to take place. This accounts for the *oxygen debt* incurred after strenuous muscular exertion.

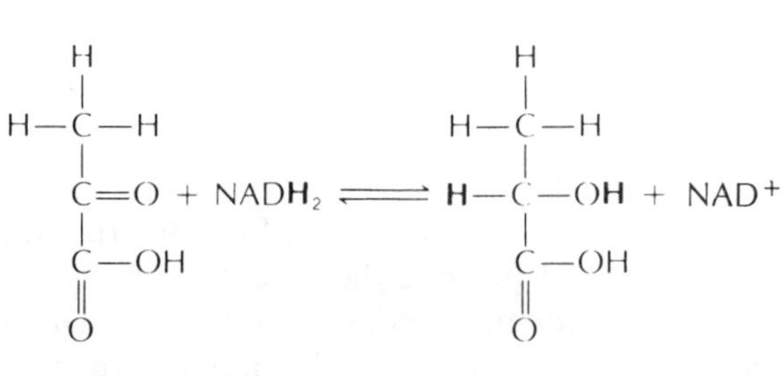

$$
\begin{array}{c}
\mathrm{H} \\
| \\
\mathrm{H{-}C{-}H} \\
| \\
\mathrm{C{=}O} \\
| \\
\mathrm{C{-}OH} \\
\| \\
\mathrm{O}
\end{array}
+ NADH_2 \rightleftharpoons
\begin{array}{c}
\mathrm{H} \\
| \\
\mathrm{H{-}C{-}H} \\
| \\
\mathrm{H{-}C{-}OH} \\
| \\
\mathrm{C{-}OH} \\
\| \\
\mathrm{O}
\end{array}
+ NAD^+
$$

Pyruvic acid   **Lactic acid**

Fig. 10-3. The interconversion between pyruvic acid and lactic acid, a critical reaction in the anaerobic recycling of NAD^+ reduced in glycolysis. This oxidation-reduction reaction is catalyzed by the enzyme lactic acid dehydrogenase, which has NAD^+ as its coenzyme.

Citric Acid Cycle

When oxygen is available, pyruvic acid is oxidized completely to CO_2 and H_2O by means of the *citric acid cycle*. This aerobic phase begins with the oxidation of pyruvic acid to activated acetic acid, or *acetyl CoA* (acetyl coenzyme A). The reaction is a complex one, requiring coenzyme A and several other coenzymes, including NAD^+. The 3-carbon pyruvic acid loses a carbon atom as a molecule of CO_2, as well as 2 hydrogen atoms, in being converted to the 2-carbon acetyl fragment attached to CoA. This crucial and irreversible reaction, which is the connecting link between glycolysis and the citric acid cycle, takes place in the mitochondria. Since 2 pyruvic acids (pyruvate) are formed from the original glucose molecule, the overall reaction can be summarized as follows:

citric acid cycle

acetyl CoA

$$\underset{(C_3)}{2\text{ pyruvate}} + 2NAD^+ + CoA \rightarrow \underset{(C_2\text{—}CoA)}{2\text{ acetyl CoA}} + \underset{(C_1)}{2CO_2} + 2NADH_2$$

The acetyl residues are now fed into the citric acid cycle and oxidized to yield carbon dioxide and pairs of hydrogen atoms (mainly as $NADH_2$). This cycle, which also takes place in the mitochondria, is a series of reactions involving about eight organic acids (Fig. 10-4). The acetyl residues are first condensed with a 4-carbon acid, oxaloacetic acid, to form the 6-carbon acid *citric acid*. (Citric acid is a tricarboxylic acid; that is, it contains three carboxyl groups.) Citric acid is then converted in a stepwise fashion back to oxaloacetic acid, grinding up the 2-carbon fragment in the process. For every glucose molecule, the reactions can be summarized as follows:

citric acid

$$\underset{(C_2)}{2\text{ acetyl CoA}} + \underset{(C_4)}{2\text{ oxaloacetic acid}} + 2H_2O \rightarrow \underset{(C_6)}{2\text{ citric acid}} + 2CoA$$

$$\underset{(C_6)}{2\text{ citric acid}} + 6NAD^+ + 2FAD + 4H_2O + 2GDP + 2P_i \rightarrow$$

$$\underset{(\mathbf{C_4})}{\mathbf{2\text{ oxaloacetic acid}}} + 6NADH_2 + 2FADH_2 + \underset{(C_1)}{4CO_2} + 2GTP$$

Note the regeneration of the key 4-carbon acceptor molecule, *oxaloacetic acid* (shown in bold), in the cycle. Note also that 1 high-energy phosphate compound is formed directly by substrate-level phosphorylation in each turn of the cycle (*two* turns per glucose molecule). The high-energy compound produced here by

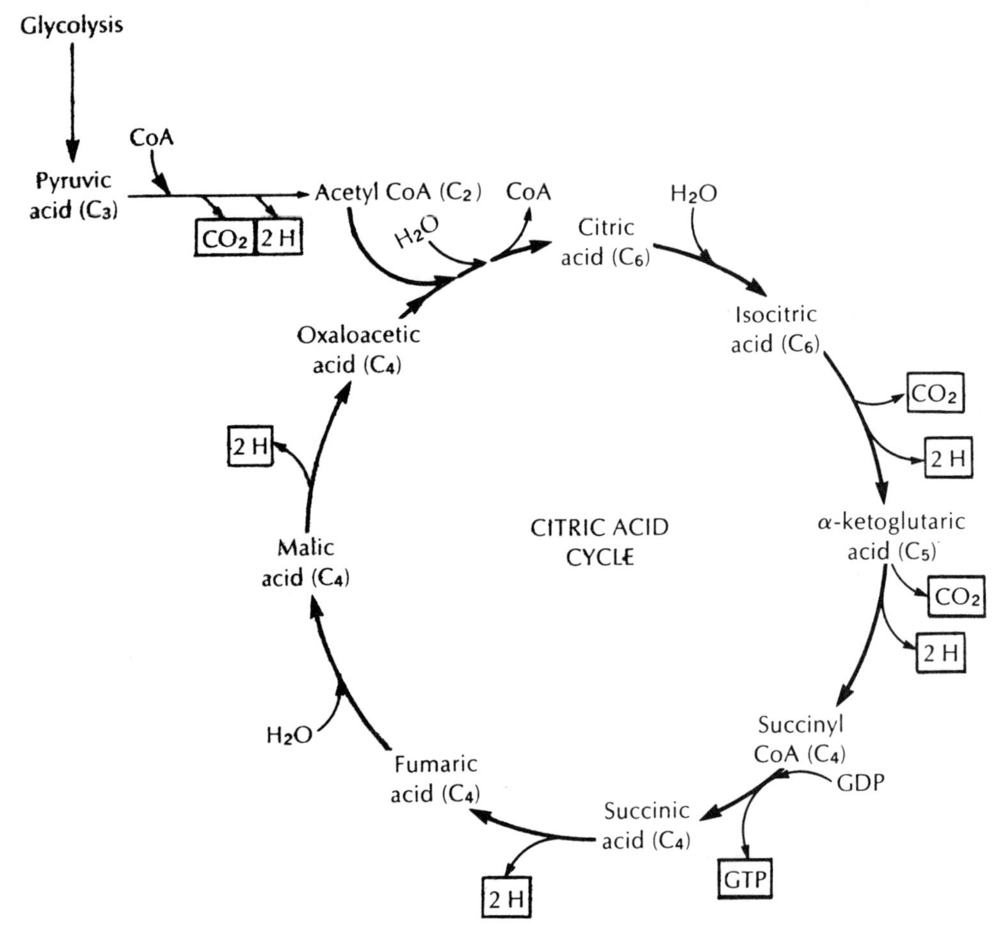

Fig. 10-4. Citric acid cycle. The reaction pathway moves in a clockwise direction, starting with the condensation of the 2-carbon acetyl CoA and the 4-carbon oxaloacetic acid, to form the 6-carbon citric acid. Two carbon atoms enter (as acetyl CoA) and 2 carbon atoms leave (as 2 CO_2). The cycle makes *two* turns per glucose molecule, producing a total of 16 H, 2 GTP (ATP), and 4 CO_2.

the conversion of succinyl CoA to succinate is another nucleotide, GTP, which is converted to ATP.

Electron Transport and Oxidative Phosphorylation

We have followed the changes in a food molecule (glucose) as it was processed in a series of stepwise reactions that take place in two compartments of the cell, firstly in the cytosol, and then in the mitochondria. During these reactions, the carbon chain of the glucose molecule is broken down (oxidized), and the free energy of its chemical bonds, in the form of its hydrogen atoms or electrons (depending on the carrier), is extracted. The total yield of liberated hydrogens at various stages of the oxidative pathway can be summed up as follows:

1. Glycolysis (glucose to 2 pyruvates) → $2NADH_2$
2. Oxidation of 2 pyruvates to 2 acetyl CoA → $2NADH_2$
3. Oxidation of 2 acetyl CoA in citric acid cycle → $6NADH_2$ and $2FADH_2$

Thus, a total of twelve pairs of hydrogen atoms are removed from each glucose molecule that is oxidized. These energetic hydrogens, mainly in the form of their electrons, will be transported down a system of enzymes and carriers, located on the inner mitochondrial membrane, which is known as the *electron transport chain,* or *respiratory chain.*

respiratory chain

At this point, some details of the functional microanatomy of mitochondria are relevant. Mitochondria *(sing.—mitochondrion)* are compartmentalized organelles in most cells, which may contain a few to hundreds of them. They are enclosed in two membranes, the *outer mitochondrial membrane* and the *inner mitochondrial membrane* (Fig. 10-5A). Both membranes are characteristic lipid bilayers with embedded protein molecules, many of them enzyme complexes. The inner membrane is thrown up into folds, called *cristae,* which greatly increase its surface area. The presence of the inner membrane results in the formation of two separated fluid subcompartments in the interior of the organelle, namely, the *intermembrane space* between the inner and outer membranes, and the central subcompartment, or *matrix,* enclosed in the folds of the inner membrane. The matrix contains most of the soluble enzymes of the mitochondrion, including the enzymes of the citric acid cycle. The inner mitochondrial membrane is the site of the two sets of components that mediate oxidative phosphorylation, namely, the electron transport chain and the F_1-F_o complexes.

inner & outer mitochondrial membranes

cristae

intermembrane space

mitochondrial matrix

The electron transport chain (Fig. 10-6). The chain includes the following components: (1) a sequence of three enzyme complexes embedded in the inner mitochondrial membrane; the latter are very large multisubunit transmembrane proteins that extend through the full width of the inner membrane, and (2) two carriers on the membrane that link the complexes; their function is to shuttle electrons between the complexes. Complex 1 on the chain is called *NADH-CoQ reductase;* Complex 2 is called *$CoQH_2$-cyto-*

electron transport chain

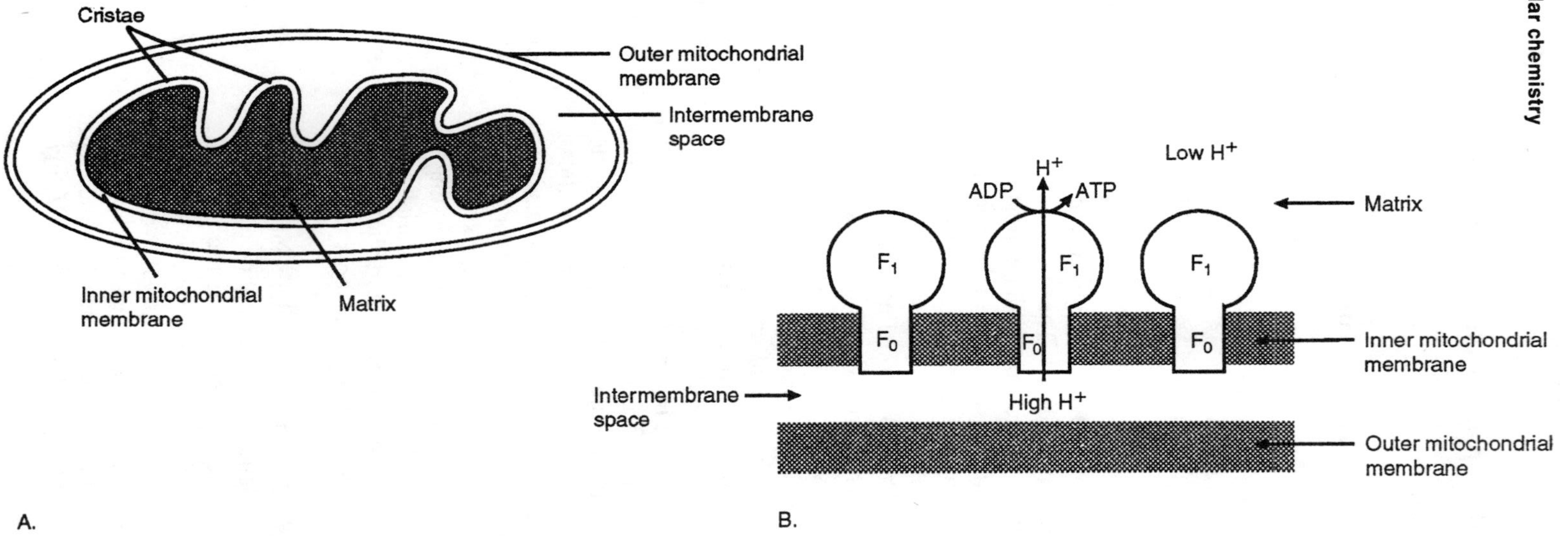

Fig. 10-5. **(A)** Schematic structure of a mitochondrion showing the compartmentation of the organelle. **(B)** Detail of the inner mitochondrial membrane with embedded F_1-F_0 complexes. As a result of proton pumping (see text), the proton (H^+) concentration in the intermembrane space is higher than in the matrix. The flow of protons through the F_1-F_0 complexes drives the synthesis of ATP from ADP.

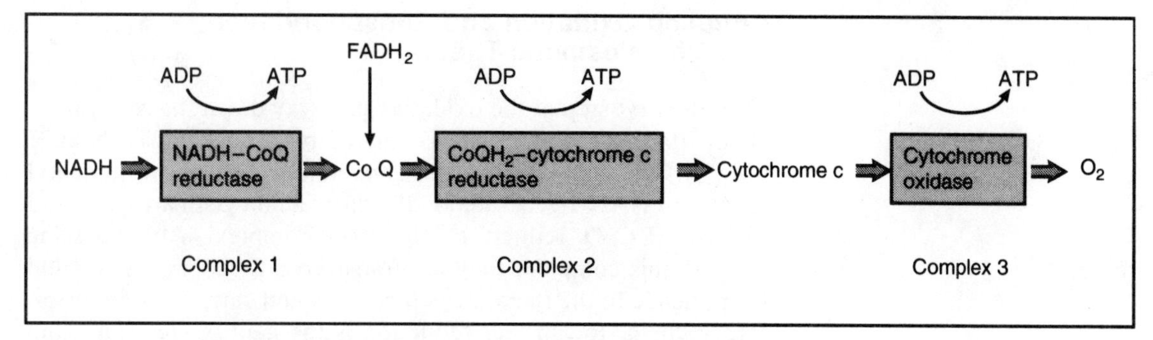

Fig. 10-6. Components of the electron transport chain (respiratory chain) on the inner mitochondrial membrane. The three complexes shown in gray are the proton pumping sites; this is where ATP is synthesized. Note that each pair of electrons carried by mitochondrial NADH passes through 3 sites, and generates 3 ATP. Each pair of hydrogens carried by $FADH_2$ bypasses the first pumping site and enters the chain just before the second pumping site, generating 2 ATP. The final electron acceptor on the electron transport chain is molecular oxygen.

chrome c reductase; Complex 3, the final complex on the chain, is *cytochrome oxidase*. The carrier between Complexes 1 and 2 is *coenzyme Q* (also called CoQ, or ubiquinone), the only non-protein component of the chain. It is a lipid substance that is soluble in the lipid core of the membrane, and presumably mobile. The carrier between Complexes 2 and 3 is *cytochrome c,* a small (12,000 d) water-soluble protein that lies on the exterior surface of the membrane. Of interest here are the several cytochromes of the respiratory chain. The iron-containing *heme* group that is present in hemoglobin is also found in the cytochromes. However, the cytochromes are evolutionarily much older than hemoglobin. Iron (Fe) is a convenient metal for electron transfer in that it can be sequentially oxidized to its ferric (Fe^{3+}) form and reduced to its ferrous (Fe^{2+}) form.

F_1-F_o complexes

The F_1-F_o complexes (Fig. 10-5B). The other important sequence of components on the inner mitochondrial membrane is the embedded mushroom-shaped F_1-F_o complexes. The F_1 'knob' of the complex, which protrudes into the matrix from the inner surface of the membrane, is an enzyme, *ATP synthase*. This catalyzes the reaction ADP + $P_i \rightarrow$ ATP, that is, the phosphorylation of ADP to form ATP. The reaction is, of course, energetically uphill, and the enzyme operates only when there is a source of energy to drive it. The F_o part of the complex serves as a transmembrane channel through which protons (H^+) cross the membrane and gain access to the catalytic F_1 unit.

ATP synthase

Coupling Oxidation and Phosphorylation: The Chemiosmotic Theory

In the first step of the oxidative pathway down the respiratory chain, the hydrogen electrons carried by mitochondrial NADH (i.e., NADH generated by the conversion of pyruvate to acetyl CoA, and NADH generated in the citric acid cycle) are delivered to NADH-CoQ reductase, the first complex. After passing through this complex, they are transferred to Coenzyme Q, and from thence to the remaining complexes and carriers of the respiratory chain. Recall that reduction is the gain of electrons, and oxidation the loss of electrons. The components of the respiratory chain are sequentially reduced and oxidized as they accept and then give up (lose) each pair of electrons, transferring them to the next component on the chain. At the end of the chain, the cytochrome oxidase complex loses the pair of electrons to an oxygen atom, and two protons ($2H^+$) are picked up from the mitochondrial matrix to form a water molecule:

$$2e^- + 2H^+ + \frac{1}{2}O_2 \rightarrow H_2O$$

In this oxidative pathway, energy is extracted in discrete packets as the hydrogen electron pairs are transferred step by step down the sequence of complexes and carriers on the respiratory chain to the final electron acceptor, oxygen, with the formation of the low-energy compound, water.

We now come to the central question of bioenergetics: *how is oxidation (the downhill transfer of electrons from oxidized substrates to molecular oxygen) coupled to phosphorylation (the uphill synthesis of ATP from ADP and P_i)?* In the 1960's, the British biochemist, Peter Mitchell, postulated a highly innovative concept, called the *chemiosmotic theory,* to answer this question, and in doing so, revolutionized our understanding of bioenergetics. He was awarded the Nobel Prize in Chemistry in 1978.

chemiosmotic theory

According to the chemiosmotic model, which has since been confirmed by experimental findings in many laboratories, each pair of electrons passing through complexes 1, 2, and 3 results in the translocation (pumping) of protons (H^+) from the matrix to the intermembrane space. The three complexes are often called 'proton pumping' sites 1, 2 and 3. The result of this activity is that the H^+ concentration in the intermembrane space is much higher than in the matrix. In other words, a *proton gradient* exists across the inner mitochondrial membrane (see Fig. 10-5B). It takes an investment of energy to establish this gradient since the protons are being pumped uphill. The energy is supplied by siphoning off the energy of electrons as they flow through the elec-

proton gradient

tron transport chain. Once free energy has been invested to establish the gradient, the energy can be recovered by allowing the protons to flow back across the inner mitochondrial membrane. This is where the F_1-F_o complexes enter the picture. The protons actually flow down their gradient through the F_o channels, and the gradient energy, which Mitchell termed the *proton-motive force*, is harnessed to drive the synthesis of ATP in the catalytic F_1 unit, the ATP synthase of the complex.

proton-motive force

The sequential arrangement of the three complexes on the inner mitochondrial membrane that are proton pumping sites is such that they operate in close association with F_1-F_o complexes. Thus, as each electron pair passes through a site, protons are pumped out, and the ATP synthase unit of the F_1-F_o complex phosphorylates one molecule of ADP to ATP. The three sites are frequently called 'phosphorylation sites' or 'coupling sites'. Electron pairs, such as those carried by mitochondrial NADH, pass through all 3 sites and generate 3 ATP per electron pair. The carrier, $FADH_2$, which is produced in the citric acid cycle, enters the electron transport chain at CoQ, bypassing the first proton pumping site; it therefore has a yield of 2 ATP per electron pair it carries. Similarly, each pair of electrons carried by the NADH's of glycolyis generates 2 ATP, rather than 3. This is because the outer mitochondrial membrane is impermeable to cytosolic NADH, and its electrons have to be carried into the mitochondria by a shuttle mechanism which loses 1 ATP in the process.

Finally, it should be noted that we owe to Mitchell a clear understanding of the meaning of *gradient energy*, and the use of this very important source of free energy by cells for metabolic, transport and other purposes (see also Chapter 9).

gradient energy

Two forms of energy are produced by oxidative phosphorylation: heat energy and the free energy that is stored in ATP. The heat is a byproduct; although it keeps us warm, it is eventually lost from the body. Heat cannot be used as a source of energy for the cellular engine. However, the energy stored in ATP is useful energy that can be put to work for the cell.

By referring to the beginning of this section, where the hydrogen atoms (and their carriers)—produced at each stage of the pathway of glucose oxidation—are listed, the yields of ATP by oxidative phosphorylation can be estimated as follows:

1. Glycolysis: 2 $NADH_2$, each yielding 2 ATP (total: 4 ATP)
2. Oxidation of pyruvate to acetyl CoA: 2 $NADH_2$, each yielding 3 ATP (total: 6 ATP)
3. Oxidation of acetyl CoA in the citric acid cycle: 2 $FADH_2$, each yielding 2 ATP, and 6 $NADH_2$, each yielding 3 ATP (total: 22 ATP)

The total yield of ATP by oxidative phosphorylation is thus 32 ATP, about 90% of all the ATP generated by the oxidation of glucose. We can now add to this the 2 ATP produced directly during glycolysis and the 2 GTP (converted to ATP) produced directly during the citric acid cycle, making a grand total of *36 ATP* formed for each completely oxidized glucose.

The complete oxidation of 1 mole of glucose (180 g) releases about 686 kcal of energy, and the energy content of 36 ATP is about 263 kcal. Thus, the overall efficiency of this oxidative pathway is 263/686, or about 38%. (This is actually a highly efficient form of energy production!) A summary of the metabolic pathways and energy yield in the complete oxidation of 1 glucose molecule is given in Table 10-1.

Table 10-1. Metabolic pathways and energy yield in the complete oxidation of 1 glucose molecule

Glycolysis

Skeleton reaction:

Glucose $\rightarrow$ 2 pyruvic acid

Overall reaction:

Glucose + 2ADP + $2NAD^+$ + $2P_i \rightarrow$ 2 pyruvic acid + 2ATP + $2NADH_2$ + $2H_2O$

Oxidation of pyruvic acid to acetyl CoA (activated acetic acid)

Skeleton reaction:

Pyruvic acid $\rightarrow$ Acetic acid + CO_2

Overall reaction:

Pyruvic acid + NAD^+ + CoA $\rightarrow$ Acetyl CoA + $NADH_2$ + CO_2

Citric acid cycle

Skeleton reaction:

Acetic acid + $2H_2O \rightarrow 2CO_2$ + 8H

Overall reaction:

Acetyl CoA + $3NAD^+$ + FAD + GDP + P_i + $2H_2O \rightarrow 2CO_2$ + $3NADH_2$ + FAD H_2 + GTP

Oxidative phosphorylation

Skeleton reaction:

2H + $\frac{1}{2}O_2 \rightarrow H_2O$

Overall reaction:

24 hydrogen protons and electrons + 6 molecules of oxygen (O_2 from air) $\rightarrow 12H_2O$ + *energy*

↙ ↘

32ATP heat

Summary of glucose oxidation

Glucose + $6O_2 \rightarrow 6CO_2$ + $6H_2O$ + 36 high-energy phosphate bonds (ATP) + heat

METABOLISM IN GENERAL

The metabolic chemistry described in this chapter is common to all cells in the body. It is controlled and integrated by complex feedback mechanisms that operate at two levels: within the cell and throughout the body as a whole. The pattern of metabolism in living systems resembles that of a planned economy. As we shall see, catabolic and anabolic pathways converge on and branch off from the reaction mainstream at various points; all pathways share common intermediate metabolites; all substrates that are catabolized for energy are eventually funneled through the electron transport—oxidative phosphorylation reaction sequence.

Catabolic Pathways

The citric acid cycle is the *final common pathway* for the catabolism of carbohydrates, fats, and proteins—whether derived from ingested food (*exogenous*, from sources outside the body) or from storage fat, storage glycogen, or body protein (*endogenous*, from sources within the body). The entry of these substrates into the pathway is shown in Fig. 10-7. Note that the nutrient molecules—fats, carbohydrates, and proteins—are first broken down enzymatically (by digestion in the alimentary tract or by intracellular digestion) to their component building blocks before they are oxidized for energy.

Lipid catabolism. Neutral fats are first hydrolyzed to fatty acids and glycerol. Glycerol is converted to glyceraldehyde 3-phosphate and enters the glycolysis pathway. Fatty acid chains are oxidized in a stepwise manner, entering the pathway as 2-carbon molecules of acetyl CoA. The acetyl CoA produced by fat (and protein) catabolism may be converted in the liver to the *ketone bodies,* acetoacetic acid, and β-hydroxybutyric acid. Many tissues use the ketone bodies produced by the liver as fuel for the citric acid cycle. However, if there is a lack of readily available glucose (as in chronic starvation) or an inability to metabolize glucose (as in diabetes mellitus), excessive breakdown of fat and protein for energy may lead to an accumulation of ketone bodies. This is commonly the cause of *metabolic acidosis* (ketosis).

ketone bodies

metabolic acidosis

Protein catabolism. Surplus protein in the diet, and body proteins that are being replaced by turnover, are hydrolyzed to amino acids and used as fuel. Amino acids are first processed by *deamination;* that is, the amino (NH_2) groups are removed. The carbon skeletons that remain are fed into the oxidation pathway by being

deamination

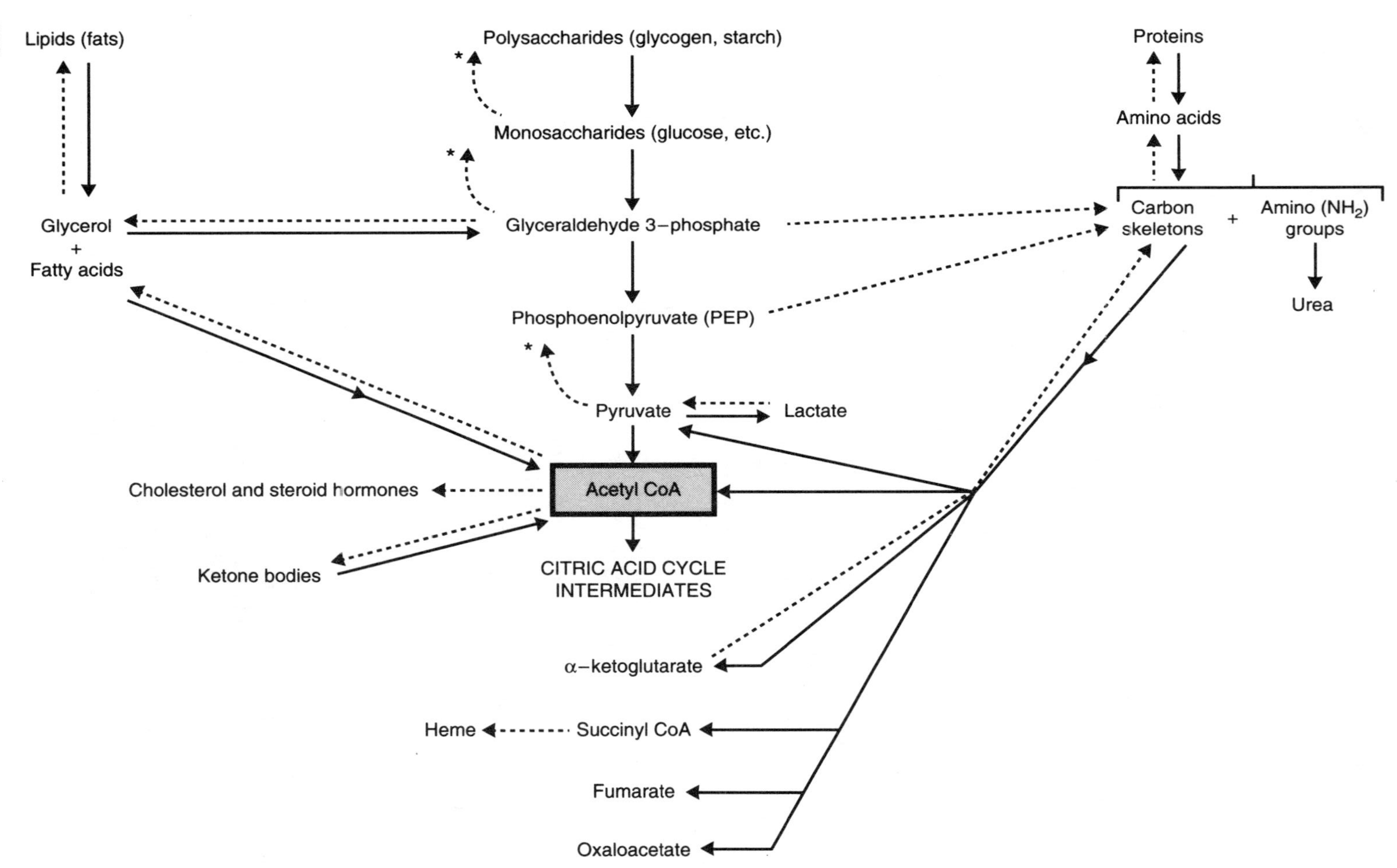

Fig. 10-7. Summary of metabolic pathways in the cell. Catabolic pathways are indicated by solid arrows; anabolic pathways by dashed arrows.

* Alternate (not directly reversible) anabolic pathways.

converted to pyruvic acid, acetyl CoA, or various intermediate acids of the citric acid cycle. Note that the NH_2 groups may be converted by liver cells to the waste product *urea*

urea

$$H_2N-\underset{\underset{O}{\|}}{C}-NH_2$$

or they may be reutilized for the biosynthesis of new amino acids.

Carbohydrate catabolism. Polysaccharides, such as starch and glycogen, and various dietary sugars are hydrolyzed to monosaccharides (for example, glucose), which converge with glycerol on the glycolysis intermediate, glyceraldehyde 3-phosphate.

Anabolic Pathways

The intermediate metabolites of the oxidative pathways also provide precursors for the synthesis of various compounds that are stored or otherwise used by the body. These pathways are indicated by dashed lines in Fig. 10-7. The major anabolic pathways can be outlined as follows:

1. Gluconeogenesis. Amino acids, lactic acid, and glycerol (but *not* fatty acids, in man) can be converted to glucose and thence to glycogen (see also Fig. 5-8). Some reaction sequences of gluconeogenesis are not simply reversed glycolysis reactions but proceed by more roundabout routes.
2. Dietary excess of protein or carbohydrate can be converted to fat and stored by the body; note the dashed lines (Fig. 10-7) connecting various substrates and various intermediates of the oxidation pathway to the pathway for the biosynthesis of fat.
3. Acetyl CoA is the precursor for cholesterol and the steroid hormones.
4. The glycolysis intermediates, glyceraldehyde 3-phosphate and phosphoenolpyruvate, and the citric acid cycle intermediate, α-ketoglutaric acid, are important precursors for the synthesis of a number of (nonessential) amino acids.
5. The citric acid cycle intermediate, succinyl CoA, is one of the precursors for the substance *heme*, the iron-containing pigment moiety of the hemoglobin molecule.

heme

Hormone Interactions

The hormone messengers secreted by endocrine cells play an important role in the feedback mechanisms that regulate cellular metabolism. Many hormones act by switching on, or off, certain

Table 10-2. Some metabolic effects of hormones

Endocrine gland	Hormone	Effects (stimulates)
Adrenal cortex	Glucocorticoids (cortisol and others)	Glycogenolysis (glycogen → glucose) Gluconeogenesis (amino acids → glucose)
Adrenal medulla	Epinephrine (adrenaline)	Glycogenolysis (glycogen → glucose) O_2 uptake, CO_2 output Pyruvate → lactate (in muscles) Lactate → pyruvate → glucose (in liver)
Anterior pituitary	Growth hormone (somatotrophin)	Lipids → fatty acids Ketone body formation Protein anabolism (amino acids → protein) Glycogen → glucose
Islets of Langerhans, pancreas	Insulin	Glucose uptake and catabolism Biosynthesis of lipids, proteins, and glycogen
	Glucagon	Glycogenolysis (glycogen → glucose) Gluconeogenesis (amino acids → glucose)
Testes	Androgens (male sex hormones)	(Increased) basal metabolic rate (BMR) Protein anabolism (amino acids → proteins)
Thyroid	Triiodothyronine (T_3) and thyroxine (T_4)	Rate of oxidative phosphorylation Heat output ("calorigenic effect") (Increased) basal metabolic rate (BMR)

sequences of the pathways dealt with in this chapter. The study of the interrelationships between endocrine and metabolic physiology is not within the scope of this introductory text. However, a brief outline of some important effects of hormones, in the context of the biochemical reactions that have been described in this chapter, is given in Table 10-2.

Metabolic Measurements (Calorimetry)

calorimeter

The potential energy of any particular food molecule can be estimated by burning a standard quantity of it in a *calorimeter* (a device for measuring heat output) and recording the amount of heat that is liberated. Thus, the potential energy content of carbohydrates or proteins, in terms of heat energy, is 4 kcal/gram, and that of fats is about 9 kcal/gram. The high caloric value of fat is due to the highly reduced state of fatty acids, that is, the large number of hydrogen atoms that can be released during oxidation of fatty acid chains. This is the reason why a high-fat diet is an effective means of maintaining body warmth in very cold climates. Likewise, adipose tissue in the body provides a concentrated, space-conserving, and efficient form of energy storage.

The metabolic rate, or energy output, of an intact animal, in terms of the heat liberated, can be most conveniently estimated by indirect methods based on the amount of oxygen consumed in a given time. As an inspection of the overall oxidation reaction of a food molecule will show (see summary of glucose oxidation, Table 10-1), oxygen consumption and heat production are closely correlated. Metabolic studies of this type employ the *respiratory quotient,* or *RQ:*

RQ

$$\mathrm{RQ} = \frac{\text{volume of } CO_2 \text{ produced}}{\text{volume of } O_2 \text{ consumed}}$$

The RQ varies with the type of food the individual is oxidizing. For example, the oxidation of 1 molecule of glucose consumes 6 O_2, and yields 6 CO_2; the RQ of a carbohydrate would therefore be 6/6, or 1. Oxidation of a fat, such as triolein, has an RQ value of 57/80, or 0.71. Protein RQ's are more difficult to measure but are considered to be about 0.80. A person on a mixed diet generally has an RQ of about 0.82 to 0.85. Under standard conditions at an RQ rate of about 0.85, the heat equivalent of 1 liter of oxygen consumed is about 4.8 kcal. If this value of 4.8 is multiplied by the number of liters of oxygen consumed during a given test period by an individual, the energy output in kilocalories per hour can be obtained. The actual metabolic rate has been found to be related to body surface area. This is determined in square meters (m^2) from a formula, using the height and weight of the individual being tested. Metabolic rate is thus measured in terms of

$$\text{kcal/m}^2 \text{ of body surface area/hour}$$

Heat output at a basal level is often expressed as the *basal metabolic rate (BMR).* The term *basal* implies that under these conditions, the body is using only a minimal amount of energy to maintain activities such as respiration, heartbeat, and normal body temperature. In this test, the oxygen uptake is measured during a given period in a resting individual about 12 hours after

BMR

a meal. The measurement is converted to kcal/m^2/hour and compared to a standard value obtained for an individual of the same sex and age. The result is reported as a percentage of the standard value; plus or minus 15% is considered within the normal range. Since the thyroid hormones have potent stimulatory effects on the overall metabolic rate, an increased BMR may indicate overactivity of the thyroid gland *(hyperthyroidism)* and a decreased BMR, underactivity *(hypothyroidism)*. However, the BMR estimation is too variable to be of value in the diagnosis of disease, and more significant tests of thyroid function have superseded it.

SDA

One interesting aspect of heat production in the body is that the absorption and distribution of digested food molecules, as such, appears to increase the metabolic rate. This phenomenon is called the *specific dynamic action (SDA)* of food. Proteins have the highest SDA value, increasing the metabolic rate (basal) by about 25%; fats and carbohydrates have a lower SDA of about 5%.

11

THE CHEMICAL STRUCTURE OF THE NUCLEIC ACIDS—DNA AND RNA

The two types of *nucleic acids* found in the cells of all living organisms are deoxyribonucleic acid (DNA) and ribonucleic acid (RNA).

nucleic acids

The *chromosomes* in the nucleus of the cell are composed mainly of DNA. Chromosomes are discrete bodies that duplicate themselves and are then passed from one cell to its descendants. The DNA contains the *genetic information* that is continuous from generation to generation of cells and organisms.

chromosomes

genetic information

Nuclear DNA is responsible for the synthesis of RNA. Most of the RNA produced in the nucleus moves out into the cytoplasm of the cell, where its primary function is the synthesis of cell proteins according to the blueprint provided by DNA. The proteins are generally of two types: the *functional proteins,* such as enzymes, transport proteins, receptor proteins, antibodies, certain hormones, and so on, and the *structural proteins,* which make up the fabric of the cell, the intercellular substances, and, in aggregate, the shape and form of the whole organism.

DNA and RNA are *macromolecules*. An idea of their size can be gained from their molecule weights. The molecular weights of DNA preparation from various sources range from 10^6 to 10^9 d; the molecular weight of RNA's range from 25 thousand to over a million d. For comparison, proteins, also considered to be macromolecules, generally weigh 10,000 to 200,000 d; glucose and water have molecular weights of 180 and 18 d, respectively.

macromolecules

Nucleic acids are long chains of repeating subunits called *nu-*

nucleotides

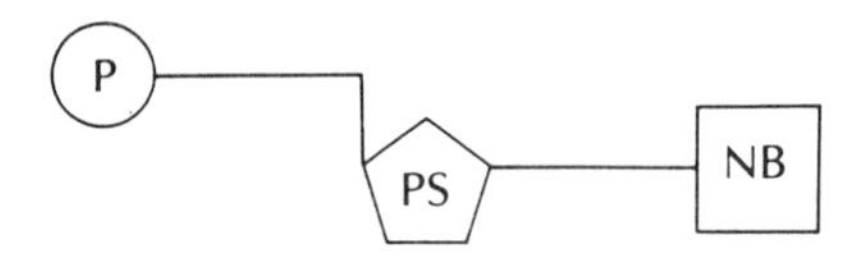

Fig. 11-1. Diagrammatic representation of a nucleotide. *P*, phosphate group; *PS*, pentose sugar; *NB*, nitrogenous base.

polynucleotide

cleotides (or more specifically, mononucleotides). A nucleic acid molecule is thus classified as a *polynucleotide* (many nucleotides). The general structure of a single nucleotide is shown diagrammatically in Fig. 11-1. It is composed of three units: a phosphate group derived from phosphoric acid (H_3PO_4), attached to a pentose sugar (5 carbons), attached to a nitrogenous base (either a pyrimidine or a purine).

NITROGENOUS BASES: PYRIMIDINES AND PURINES

nitrogenous base

The *nitrogenous bases* are organic heterocyclic compounds composed of carbon nitrogen, hydrogen, and in some cases, oxygen (see Chapter 4).

pyrimidine

cytosine
thymine
uracil

The compound pyrimidine from which the *pyrimidine bases* in nucleic acids are derived, is a six-sided ring structure formed by 2 nitrogen and 4 carbon atoms. Two different pyrimidine bases are found in DNA: *cytosine* and *thymine* (Fig. 11-2, *A*). The two pyrimidines in RNA are cytosine and *uracil*. The structure of uracil is very similar to that of thymine; uracil differs only in the absence of the CH_3 group. In nucleic acid chemistry, these three bases are conventionally symbolized by the capital letters C, T, and U.

purine

adenine
guanine

The *purine bases* in nucleic acids are derived from a parent compound that contains a pyrimidine ring fused to a five-sided ring (called an imidazole ring). Two purine bases are present in both DNA and RNA. They are *adenine* and *guanine,* symbolized by the letters A and G (Fig. 11-2, *B*).

THE PENTOSE SUGARS: RIBOSE AND DEOXYRIBOSE

pentose
D-ribose
2-deoxy-D-ribose

A *pentose* is a simple sugar containing 5 carbon atoms. The pentose found in RNA is D-*ribose* ($C_5H_{10}O_5$); that in DNA is 2-*deoxy*-D-*ribose* ($C_5H_{10}O_4$). As is evident from the empirical for-

The chemical structure of the nucleic acids—DNA and RNA

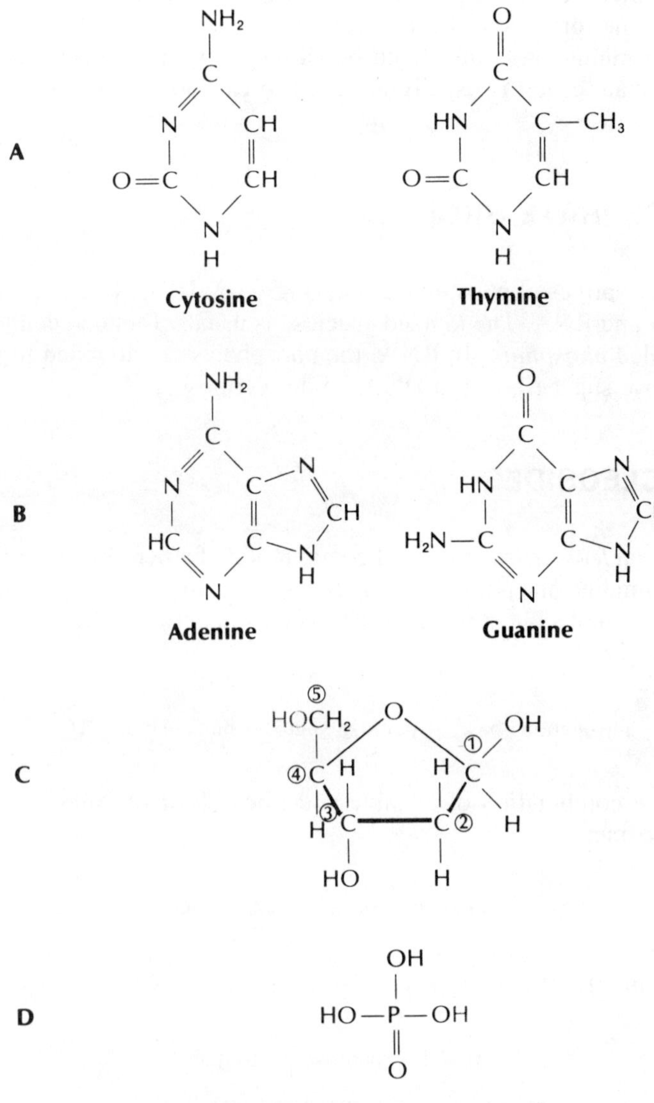

A

Cytosine **Thymine**

B

Adenine **Guanine**

C

D

Fig. 11-2. Structural formulas for DNA nucleotide components. **A,** Pyrimidine bases; **B,** purine bases; **C,** pentose sugar (deoxyribose); and **D,** phosphoric acid (phosphate). The five carbons of the pentose sugar are numbered (conventionally as 1′ to 5′). In nucleotide formation, purine or pyrimidine bases link to carbon 1′ of the pentose sugar and phosphate may link either to carbon 3′ or carbon 5′.

mulas, the two sugars differ in that 2-deoxy-D-ribose contains 1 less oxygen atom than ribose. In nucleic acids, the pentose sugars are in the form of five-membered ring structures (Fig. 11-2, *C*). Note that the 5 carbon atoms of the pentoses are numbered 1′ to 5′. In the formation of a nucleotide, as we shall see, the purine or pyrimidine base links to carbon atom 1′ of the pentose and the phosphate group is usually attached to carbon atoms 3′ or 5′ of the pentose.

PHOSPHATE GROUP

phosphate

Derivatives of phosphoric acid (Fig. 11-2, *D*) are found in both DNA and RNA. The ionized species, as usually found at cell pH, is called *phosphate*. In RNA, the phosphates are attached to the pentose sugar ribose; in DNA, to deoxyribose.

NUCLEOSIDES

nucleoside

A *nucleoside* is a chemical combination of a pentose sugar and a pyrimidine or a purine base. In the attachment of the two components, a molecule of water is removed from between the base and the sugar:

$$\text{nitrogenous base} + \text{pentose sugar} \rightarrow \text{nucleoside} + H_2O$$

The combination of adenine and ribose forms the nucleoside adenosine:

$$\text{adenine} + \text{ribose} \rightarrow \text{adenosine}$$

Similarly, the other bases also form nucleosides:

$$\text{uracil} + \text{pentose} \rightarrow \text{uridine}$$

$$\text{thymine} + \text{pentose} \rightarrow \text{thymidine}$$

$$\text{cytosine} + \text{pentose} \rightarrow \text{cytidine}$$

$$\text{guanine} + \text{pentose} \rightarrow \text{guanosine}$$

Nucleosides containing ribose are called ribonucleosides. Those containing deoxyribose are called deoxyribonucleosides.

NUCLEOTIDES

When the sugar portion of a nucleoside combines chemically with phosphoric acid, a *nucleotide* is formed (Fig. 11-1):

nucleotide

$$\text{nucleoside} + \text{phosphoric acid} \rightarrow \text{nucleotide} + H_2O$$

Nucleotides are phosphoric acid esters of nucleosides. These compounds are the building blocks of nucleic acids.

Nucleotides may be called acids (adenylic acid, guanylic acid, and so forth) because their phosphate groups produce hydrogen ions, or they may be designated as the specific nucleoside phosphate (adenosine phosphate, guanosine phosphate, and so forth). The four nucleotides found in DNA are combinations of adenine, guanine, cytosine, or thymine with deoxyribose and phosphate. The four nucleotides found in RNA are combinations of adenine, guanine, cytosine, or uracil with ribose and phosphate.

Biological Roles of Nucleotides

In addition to being the building blocks of nucleic acids, nucleotides containing one, two, or three phosphate groups (mono-, di-, and triphosphates) subserve extremely important functions in all living cells, from bacteria to man. Foremost among these is the high-energy storage compound, adenosine triphosphate (ATP), discussed in Chapter 10. ATP is a combination of the purine base adenine, the pentose sugar ribose, and three phosphate groups:

ATP

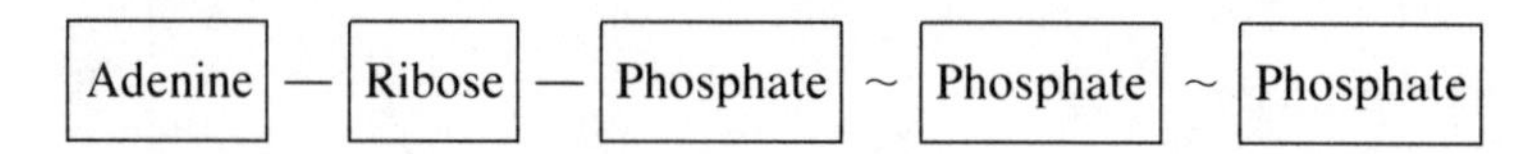

The chemical structure of the most common form of this nucleotide is adenosine 5′-triphosphate (Fig. 11-3). The energy-rich bonds in ATP are the last two phosphorus-oxygen bonds, P—O—P. The terminal phosphate bond can be hydrolyzed enzymatically, releasing 7.3 kcal of energy, inorganic phosphate (P_i), and the nucleotide *adenosine 5′-diphosphate (ADP)* (Fig. 11-3):

ADP

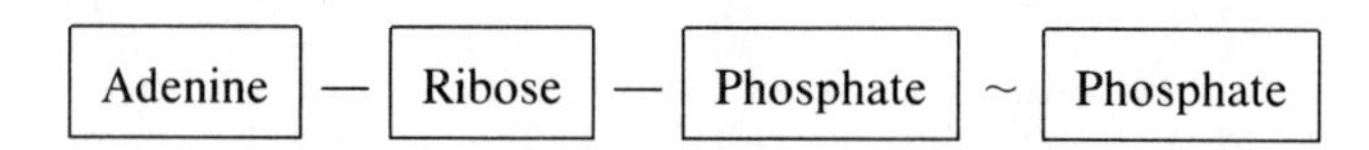

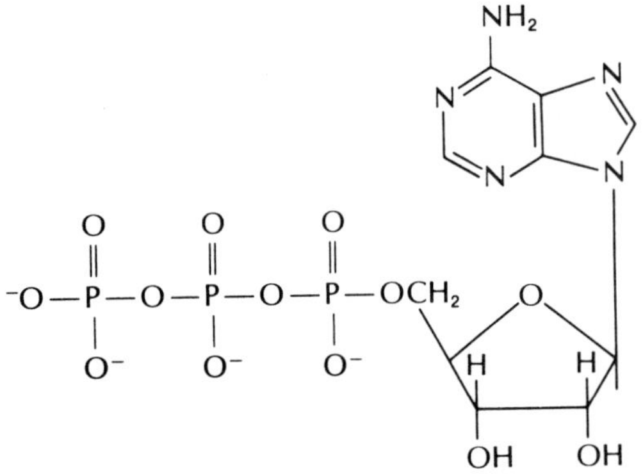

Adenosine 5′-triphosphate (ATP)

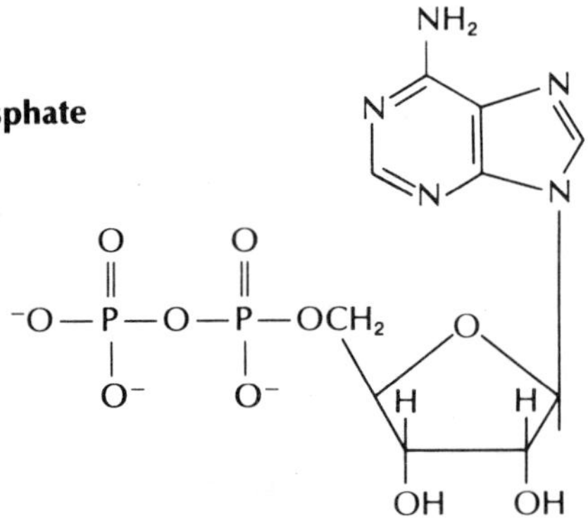

Adenosine 5′-diphosphate (ADP)

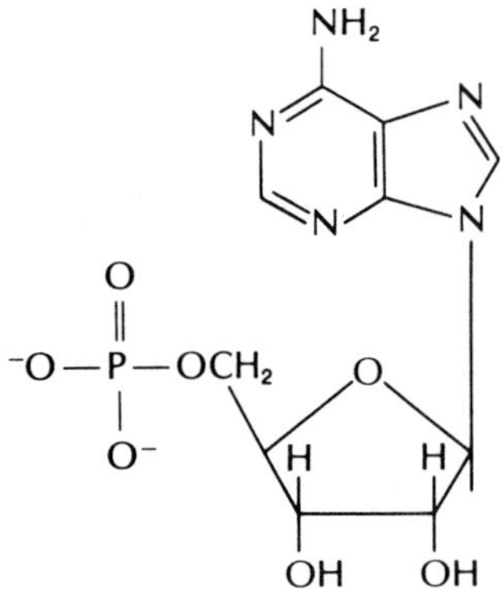

Adenosine 5′-monophosphate (AMP)

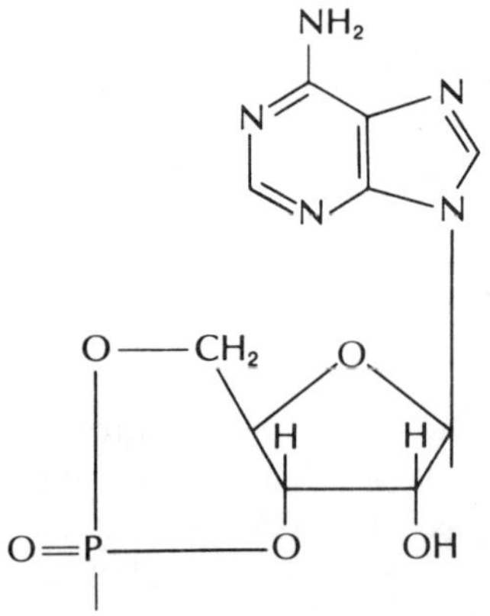

Cyclic 3′, 5′-adenosine monophosphate (cyclic AMP)

Fig. 11-3. Adenosine phosphates. At the pH of cell fluid, these compounds are negatively charged anions. The four nucleotides shown contain ribose.

The terminal phosphate bond of ADP may likewise be hydrolyzed to yield the same amount of energy, inorganic phosphate, and the nucleotide *adenosine 5′-monophosphate (AMP)* (Fig. 11-3):

AMP

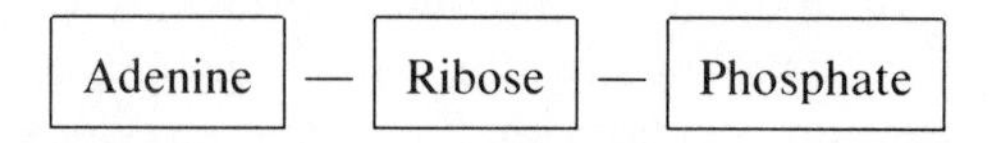

Another form of AMP found in cells is *cyclic adenosine monophosphate (cyclic AMP)* (Fig. 11-3):

cyclic AMP

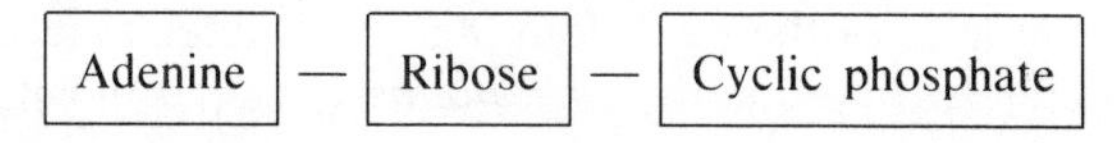

This compound differs from AMP in that the phosphate group is arranged in a ring (cyclic phosphate), sharing bonds with *two* carbon atoms (3′ and 5′) of the ribose molecule. Cyclic AMP is a vital part of the cellular response to hormonal signals; its role as the "second messenger" is described in Chapter 8.

Other nucleoside triphosphates, such as guanosine triphosphate (GTP), uridine triphosphate (UTP), and cytidine triphosphate (CTP), also function as sources of chemical energy in certain metabolic pathways and as precursors in nucleic acid biosynthesis.

The adenosine phosphates combine chemically with *vitamins* of the *B complex* to form several important *coenzymes* of the cell (see Chapter 10). Among these are:

vitamin B complex
coenzymes

1. *NAD* (nicotinamide adenine dinucleotide), a combination of AMP and a nucleotide of nicotinamide (also known as the B vitamin, *niacin*).
2. *FAD* (flavin adenine dinucleotide). This compound is a combination of ADP and a derivative of *riboflavin* (vitamin B_2). NAD and FAD are coenzymes that carry the hydrogen atoms (electrons and protons) given off by metabolites during glycolysis and the citric acid cycle to the electron transport chain in mitochondria. Here the hydrogen atoms eventually combine with oxygen to form H_2O and energy (as ATP).
3. *Coenzyme A*. This coenzyme is responsible for the formation of activated acetic acid (acetyl-CoA) from pyruvic acid, one of the most important reactions in carbohydrate and fat metabolism. Coenzyme A is a combination of ATP and the B vitamin, pantothenic acid *(pantothenate)*.

NAD
niacin
FAD
riboflavin
coenzyme A
pantothenate

The coenzymes listed above act mainly in conjunction with a group of cellular enzymes called *dehydrogenases*.

FORMATION OF THE POLYNUCLEOTIDE CHAIN: POLYMERIZATION

Thousands of individual mononucleotides (containing *one* phosphate group) link together to form a linear *polynucleotide* strand. The reaction is catalyzed by specific enzymes, DNA and RNA *polymerases.* DNA is formed from component deoxyribonucleotides; RNA is formed from component ribonucleotides. Phosphodiester bridges, running from carbon 3′ of one pentose sugar to carbon 5′ of the next, join the nucleotides together (Fig. 11-4). Polynucleotide chains have *polarity* (two opposite ends); by convention, a chain is considered to start with a phosphate group at carbon 5′ that is not linked to another nucleotide and terminate with a free hydroxyl (OH) group at carbon 3′. This pentose-phosphate-pentose-phosphate sequence forms the *backbone* of the polynucleotide chain. The pentose-phosphate linkages are made as each pentose gives up 1 hydroxyl group (OH) and each phosphate gives up 1 hydrogen atom to form 1 molecule of H_2O, which is released in the reaction as the bond is formed:

DNA and RNA polymerases

backbone

$$\text{mononucleotides} \rightarrow \text{polynucleotide} + H_2O\text{'s}$$

STRUCTURE OF THE DNA MOLECULE: THE DOUBLE HELIX

In 1953, Watson and Crick published their conclusions from investigations into the chemical nature of the DNA molecule. They showed that it consisted of two very long, single polynucleotide strands wrapped around each other in the form of a *double helix*. The structure is analogous to a coiled ladder; the outsides of the ladder are formed by the pentose-phosphate backbones, with the component purine and pyrimidine bases of each strand facing inward toward each other and interacting, thus forming the rungs of the ladder and holding the strands together. Note that the pentose-phosphate groups in the two polynucleotide chains shown in Fig. 11-4 run in opposite directions; that is, they have opposite polarity. This is a characteristic feature of double-stranded DNA. The two chains forming a DNA double helix are said to be *antiparallel;* in Fig. 11-4, the chain on the left side of the diagram starts at the top with an unlinked phosphate group at carbon 5′ of the pentose, and the chain on the right starts at the bottom.

double helix

In the DNA molecule, each base on one chain forms bonds with the opposite base on the second chain; the bonds between

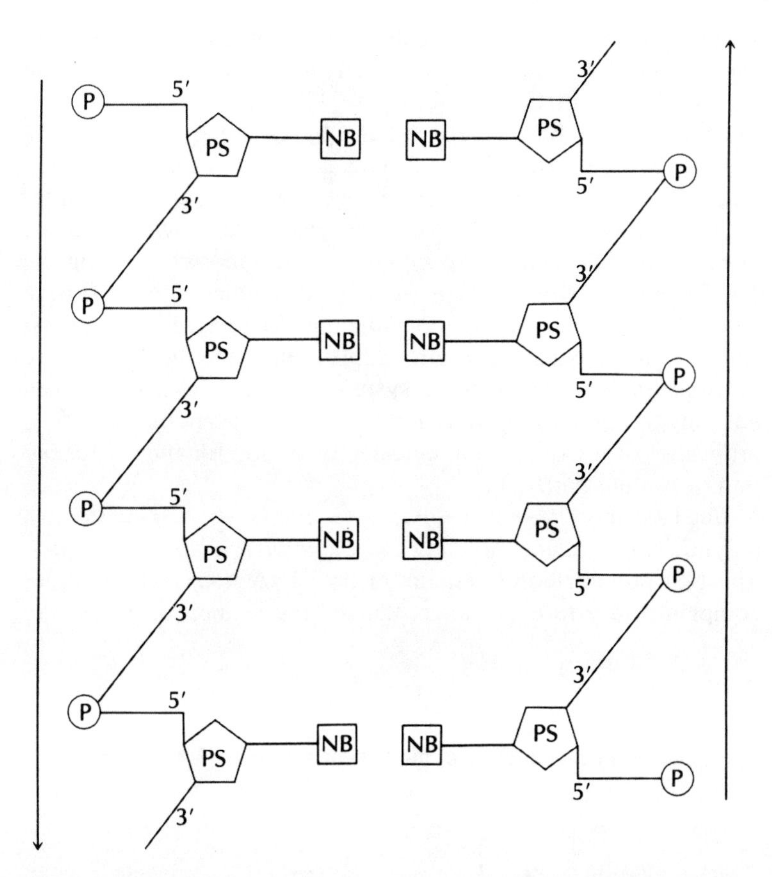

Fig. 11-4. Two polynucleotide chains shown diagrammatically. Compare with single nucleotide in Fig. 11-1 (*NB*, nitrogenous base—purine or pyrimidine; *PS*, pentose sugar; *P*, phosphate group). Note that the phosphate groups form bridges between carbon atoms 3′ and 5′ of two adjoining pentose sugars. By convention, the polynucleotide chain begins where there is an unlinked Ⓟ at the 5′ position. The beginning of the chain on the left side would thus be at the top, whereas the chain on the right would start at the bottom. In DNA, the base on one chain would be linked to the opposite base on the second chain by hydrogen bonds (see text).

these base pairs are dictated by the size and chemical structure of the bases. Purine bases (adenine and guanine) are considerably larger than pyrimidines (thymine and cytosine) (Fig. 11-2). The distance between the two polynucleotide chains is such that every *base pair* consists of a purine on one strand and a pyrimidine on the other—two purines would be too wide; two pyrimidines would be too narrow. Furthermore, the chemical structures of the bases

base pair

dictate that the only possible chemical combinations are between adenine and thymine (A—T or T—A) and between guanine and cytosine (G—C or C—G).

A—T
G—C
hydrogen bonds

The chemical bonds between the bases are *hydrogen bonds* formed by the sharing of a hydrogen atom between two strongly electronegative atoms such as nitrogen or oxygen (see Chapter 2). The hydrogen bonds between the bases are shown by dotted lines in Fig. 11-5. There are *two* hydrogen bonds connecting the A—T base pair, whereas *three* such bonds connect the C—G base pair. Hydrogen bonds are noncovalent, low-energy bonds that can be easily broken. Their strength is in their numbers. They are of great importance in living systems, since they are involved not only in nucleic acid function but are also responsible for the properties of water as a biological solvent and for the conformation of protein molecules.

complementary bases

The base pairs, A—T, T—A, C—G, and G—C, linking the two polynucleotide chains are said to be *complementary base pairs*. The two polynucleotide chains in the DNA helix are therefore complementary to each other; that is, the sequence of bases in

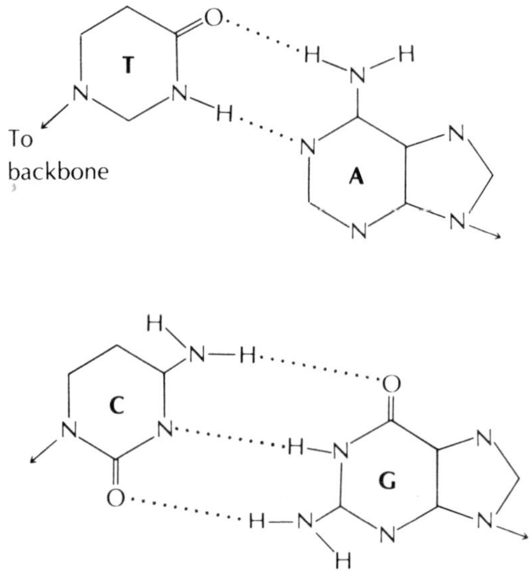

The chemical structure of the nucleic acids—DNA and RNA

Fig. 11-5. Hydrogen bonding of base pairs. Compare with structural formulas in Fig. 11-2.

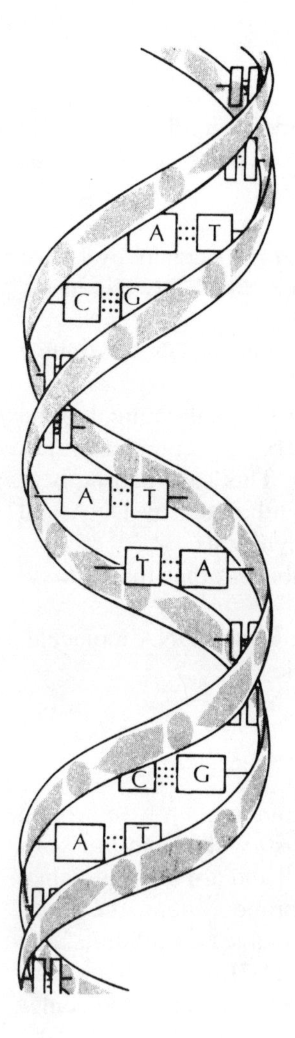

Fig. 11-6. DNA double helix. The two polynucleotide strands are antiparallel and complementary to each other. The sequence of bases in one chain determines the sequence of bases in the other. The angles of the bonds between the backbone components cause the double-stranded molecule to twist into a double helix.

one chain determines the sequence of bases in the complementary chain (Fig. 11-6). Every possible sequence of base pairs is present in a single DNA molecule which, in a human cell, may consist of over a million base pairs. The sequence of purines and pyrimidines in the DNA molecule constitutes the *genetic code* (see Chapter 12), whereas the pentose-phosphate groups provide the structural framework of the molecule.

STRUCTURE OF RNA

RNA differs from DNA chemically and in its distribution in the cell. The differences between the two types of nucleic acids are:

1. Most RNA, although synthesized in the nucleus by DNA, is found in the cytoplasm of the cell. Most DNA is in the nucleus.
2. RNA molecules are mainly single stranded. DNA is double stranded.
3. RNA does not contain the pyrimidine base thymine, but the pyrimidine *uracil* instead. In RNA, the base complementary to *adenine* is *uracil* (A—U, U—A). This base pairing is of importance when RNA is being synthesized by DNA and when RNA is involved in protein synthesis.
4. RNA contains the pentose sugar ribose rather than deoxyribose.
5. There is only one general structure for the DNA molecule. There are three distinct RNA species.

Ribosomal RNA

rRNA
ribosomes
endoplasmic reticulum
nucleolus

Ribosomal RNA *(rRNA)* is found in combination with protein in the minute round particles called *ribosomes,* which are scattered throughout the cytoplasm of the cell and are often attached to the surfaces of the intracellular membrane system, the *endoplasmic reticulum*. Mammalian ribosomes have two subunits, one of which is smaller than the other (Fig. 11-7). The subunits are composed of a total of over 50 protein molecules and 4 molecules of single-stranded rRNA. Ribosomal RNA makes up about 80% of the total RNA of the cell. It is synthesized on special regions of chromosomal DNA that are concentrated in the *nucleolus,* a small densely staining area in the nucleus of the cell.

Messenger RNA

mRNA

Messenger RNA *(mRNA)* strands exhibit considerable differences in length, with molecular masses of about 500,000 to 4 million d. Only about 5% of the total cellular content of RNA is mRNA. There is evidence that the half-life of this species of RNA may vary from very short to very long; one variety of bacterial mRNA is enzymatically broken down within a few minutes of its appearance in the cell. As we shall see in Chapter 12, the sequence

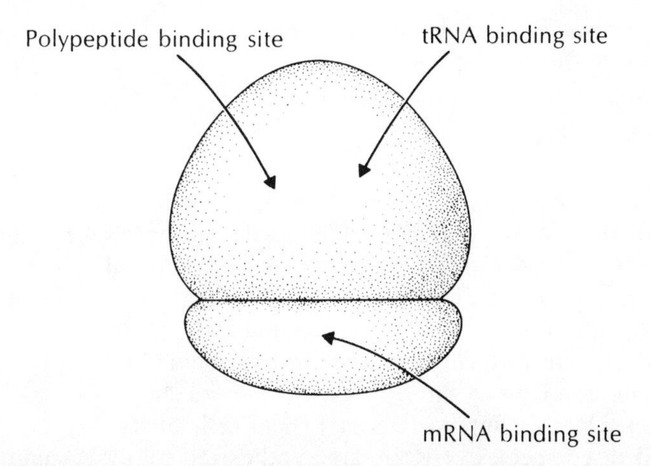

Fig. 11-7. Structure of a ribosome. This organelle is made up of protein and rRNA. The binding site for mRNA is located on the smaller of the two subunits, and the binding sites for tRNA and the lengthening polypeptide chain are located on the larger ribosomal subunit.

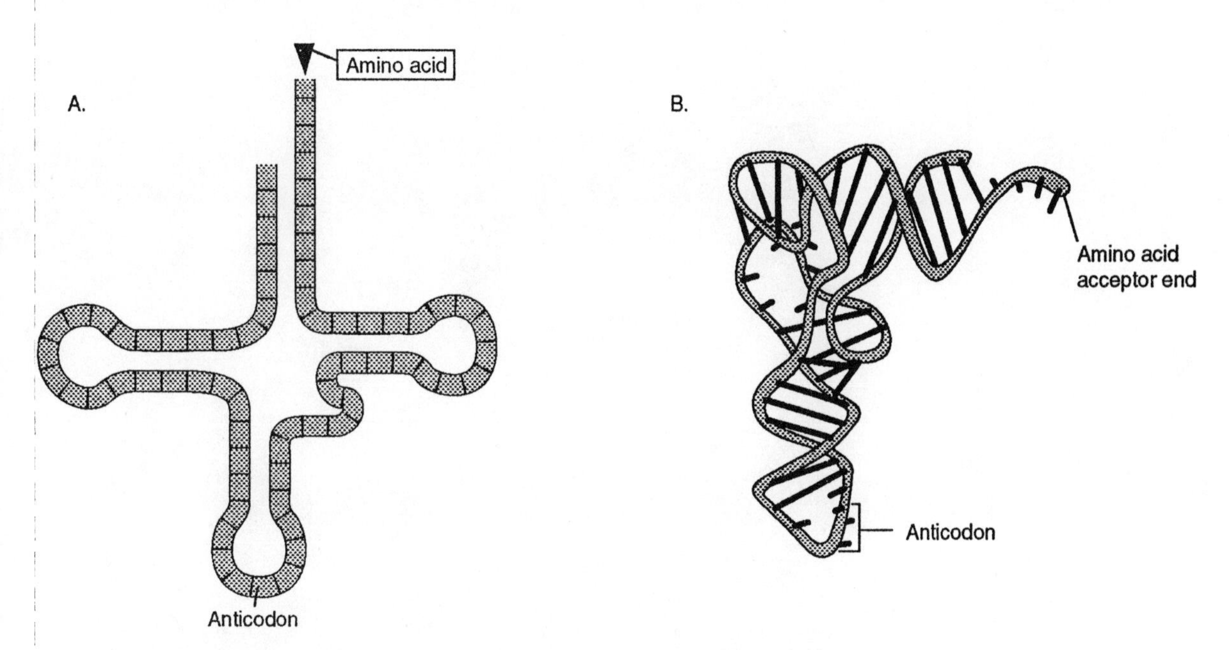

Fig. 11-8. Two diagrammatic representations of the tRNA molecule showing the anticodon end and the amino acid acceptor end. **A.** The two-dimensional cloverleaf folding pattern; **B.** The three-dimensional L-shaped structure.

of bases in mRNA molecules is complementary to the bases that constitute the genetic code.

Transfer RNA

tRNA

Transfer RNA *(tRNA)* is the smallest of the RNA species, containing only about 73 to 93 nucleotides. Transfer RNA molecules contain a comparatively large number of unusual, chemically modified bases which are mainly methylated derivatives of the four standard RNA bases, A, G, U and C. Some of the bases in tRNA pair up to form double helical segments in the molecule. Structural models of the tRNA molecule are shown in Fig. 11-8. Transfer RNA comprises about 15% of cellular RNA.

All three species of RNA are synthesized by DNA and each appears to be made on different portions of the DNA in the nucleus. The only one of the three that is imprinted with the genetic message of DNA for protein synthesis is *messenger* RNA, as its name implies. The functions of the nucleic acids, namely, DNA replication, RNA synthesis, and protein synthesis, are discussed in Chapter 12.

12

FUNCTION OF NUCLEIC ACIDS

Nucleic acids are unique biologic macromolecules necessary for the continued existence of all forms of life on earth. The *genetic information* contained in the DNA molecule is essentially a set of coded instructions for the synthesis of proteins by living cells. DNA molecules have the ability to reproduce themselves by the process of replication, thus ensuring the transmission of genetic information from one generation to the next. DNA also synthesizes RNA, and RNA in turn is responsible for the synthesis of proteins. In this way, the nucleic acids determine the ultimate form and function of all living organisms.

genetic information

REPLICATION OF DNA

The DNA molecule is a *template,* that is, a master pattern or blueprint. DNA synthesizes replicas of itself by using its own structure as a template. *Replication* is not quite the same as duplication. A duplicate is simply an exact copy of an original; a replica is a newly created structure made by using the original as a model or guide.

template

replication

Each *chromosome* in the nucleus of the cell (there are 46 in human cells) is evidently made up of a single very long double helical DNA molecule in combination with a scaffolding of basic proteins, called *histones*. At the end of interphase, just before a cell divides, each of its chromosomes appears to split into an identical pair of *chromatids*. The chromatids are replicas of the parent chromosome; they are produced by the replication of the DNA molecule in the parent chromosome. During *mitosis,* the chromatids separate and one member of each pair migrates to one of the two new daughter cells. By this mechanism, a faithful rep-

chromosomes

histones

mitosis

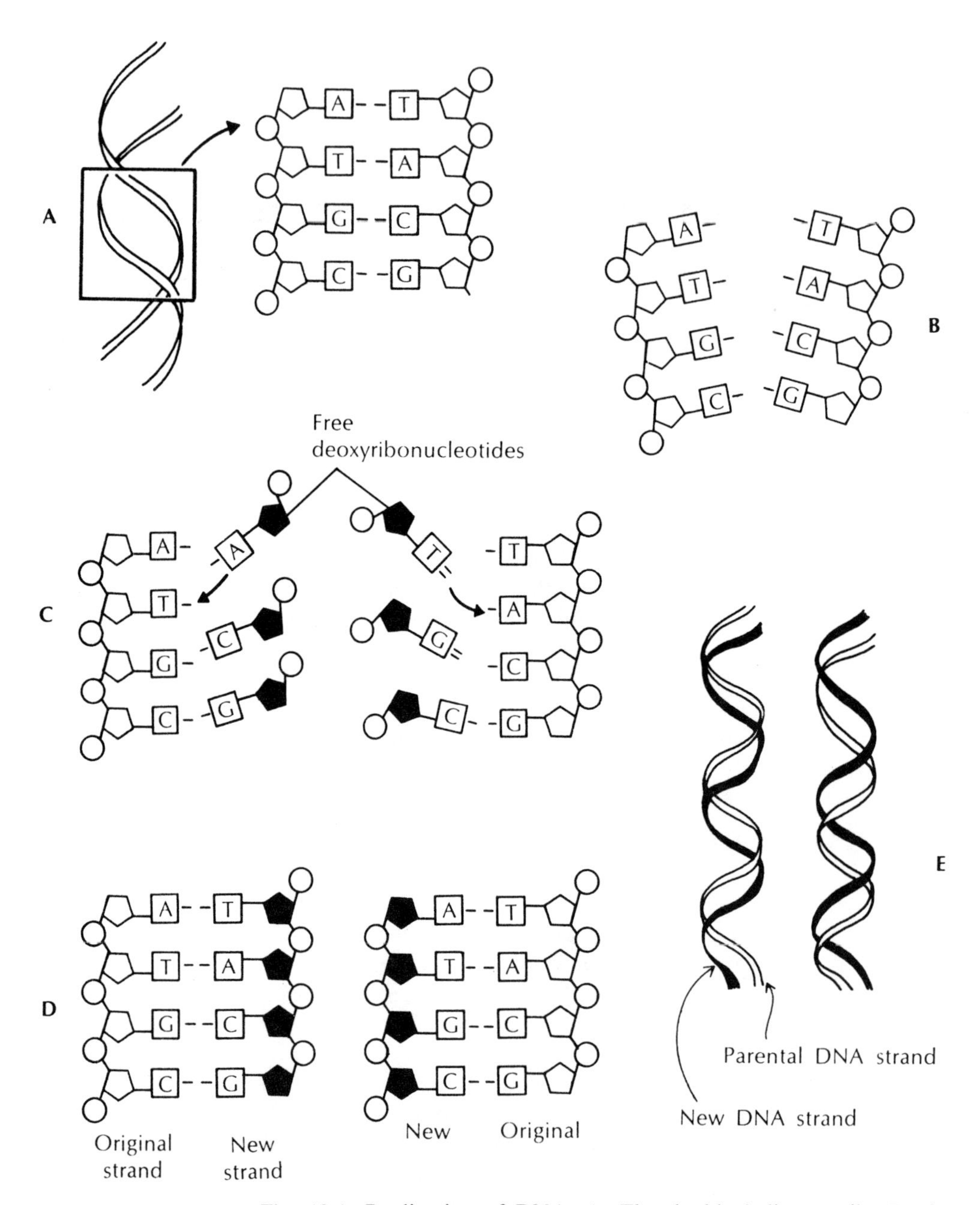

Fig. 12-1. Replication of DNA. **A,** The double helix uncoils; **B,** the strands separate; **C,** free nucleotides pair with complementary bases exposed on the DNA strands; **D,** the nucleotides polymerize and form new polynucleotide chains; and **E,** the two double-stranded daughter DNA molecules rewind into new double helices. The replicated chains are complementary and antiparallel to the parent (template) chains.

lica of the genetic information contained in the original cell is passed on to its descendants.

The two strands of DNA in the double helix replicate by an orderly sequence of steps. The two polynucleotide chains in the DNA molecule unwind (Fig. 12-1, *A*). The weak hydrogen bonds connecting the complementary base pairs on each strand are broken, and the strands separate (Fig. 12-1, *B*). Each single strand now has an exposed row of bases that serves as a template. The bases of free deoxyribonucleotides (monomers) form hydrogen bonds with these exposed bases (Fig. 12-1, *C*). The template strand dictates the sequence in which the free nucleotides are assembled; that is, the bonds can only be made between complementary base pairs (A—T, T—A, C—G, G—C). A polynucleotide backbone for the newly assembled nucleotides is formed by linkages between phosphates and deoxyriboses in the $5' \rightarrow 3'$ direction (Fig. 12-1, *D*). The two double-stranded daughter DNA molecules rewind into new double helices (Fig. 12-1, *E*). Each daughter molecule contains one newly replicated strand and one strand derived from the parent template molecule. The latter combination is termed *semiconservative replication*. The two new molecules of DNA are identical with each other and with the original parent molecule (Fig. 12-2). The complex chemistry of replication is catalyzed by several *DNA polymerases* and *DNA ligase*.

semiconservative replication

DNA polymerases
DNA ligase

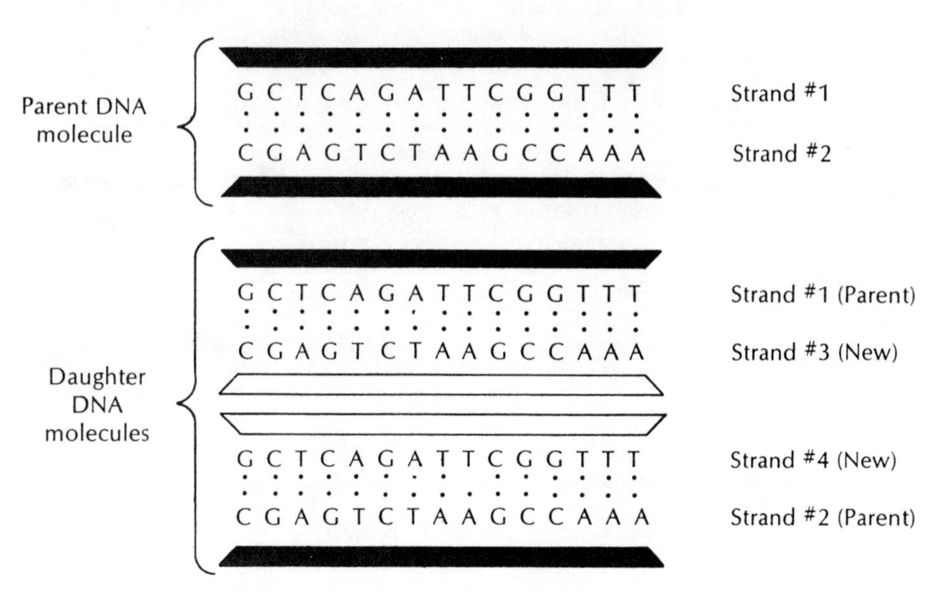

Fig. 12-2. Comparison of parent and daughter molecules of DNA.

TEMPLATE SYNTHESIS OF RNA: TRANSCRIPTION

transcription

All RNA of the cell is made by template synthesis on the DNA molecule, according to base-pairing rules. The process is called *transcription*. Transcription differs from replication in several ways, as noted in Table 12-1. However, the two processes resemble each other in that the newly synthesized RNA is complementary and antiparallel to the DNA template strand.

RNA polymerases

The three types of RNA are synthesized on different portions of the DNA molecule. The synthesis is catalyzed by three *RNA polymerases*. During transcription, only one strand of the double-stranded DNA molecule is used as a template; this strand is called the *coding strand*. The sections of DNA that transcribe *messenger RNA* contain the vital genetic "instructions" for protein synthesis.

coding strand
mRNA

Fig. 12-3 shows the biosynthesis of RNA by DNA. A portion of the DNA helix unwinds, and the two strands separate. The exposed bases on the coding strand form hydrogen bonds with complementary bases on free ribonucleotides. Ribonucleotides containing *uracil* pair with exposed *adenine* bases on the DNA template (U-A). The sequence of bases on the template molecule determines the sequence of complementary bases on the newly synthesized RNA strand. Phosphate-ribose linkages form the backbone of the polyribonucleotide. For a short time, a *hybrid molecule* exists, composed of one strand of RNA and one strand

hybrid RNA-DNA

Table 12-1. Comparison of replication and transcription

	Replication	Transcription
Fate of template DNA	Each strand ends up as part of a new DNA molecule	Returns to original double helical conformation
Nucleotide units used for the synthesis	Deoxyribonucleotides	Ribonucleotides
Formation of the polynucleotide chain	Catalyzed by DNA polymerases and DNA ligase	Catalyzed by RNA polymerases
Products formed	DNA replicas	mRNA, rRNA, tRNA molecules
Occurrence	Prior to mitosis (cell division)	Continuous process
Fate of products	Passed to daughter cells; present only in the nucleus	Leaves nucleus after synthesis; mainly found in cytoplasm

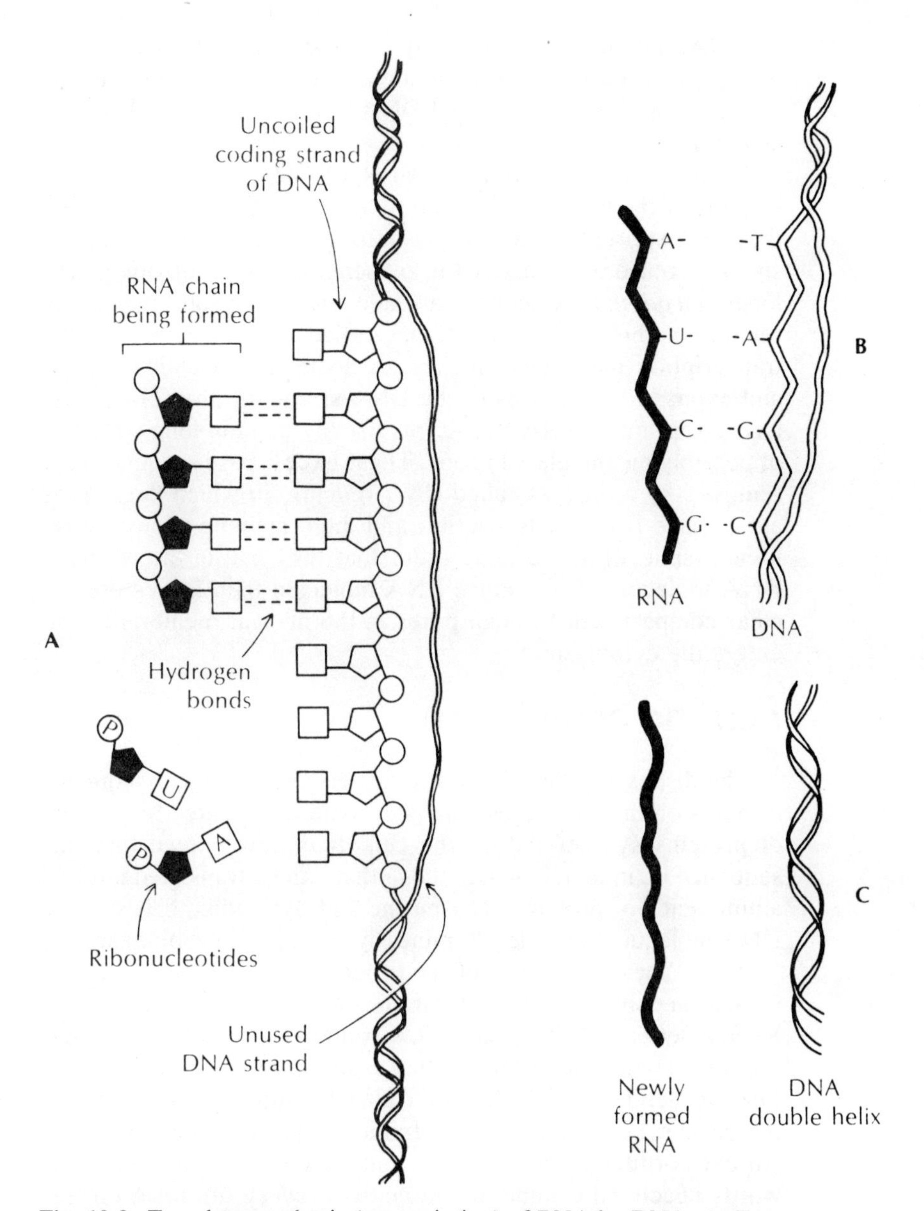

Fig. 12-3. Template synthesis (transcription) of RNA by DNA. **A,** Formation of the polyribonucleotide chain by base pairing with the coding strand of DNA. Nucleotide structure of uncoiled portion of coding strand is shown. **B,** The newly synthesized RNA strand detaches from the template. **C,** The DNA strands rewind into a double helix. After processing, the RNA leaves the nucleus and enters the cytoplasm of the cell.

of DNA. The newly transcribed single-stranded RNA molecule is then released from the template and passes from the nucleus to the cytoplasm. The original DNA strands reunite and coil up as before into a double helix.

Most of the newly-transcribed RNA molecules made in the nucleus of the cell are subjected to an 'editing' procedure before they are exported to the cytoplasm to function in protein synthesis. The reason for this is the presence in DNA of often very long *noncoding* base sequences, called *introns*. (Note, the *coding* regions of the DNA template are called *exons*.) However, the transcription mechanism makes no distinction between introns and exons; all sequences of the DNA strand are transcribed. As a result, primary RNA transcripts usually contain long stretches of possibly meaningless introns. These excess base sequences are removed by a process called *RNA splicing,* in which the introns are excised from the transcript, and the remaining exons are rejoined (spliced) in the same order that they had in the template DNA molecule. The mature RNA molecule then leaves the nuclear compartment through pores in the nuclear membrane, and enters the cytoplasm.

noncoding introns
coding exons

RNA splicing

GENETIC CODE

The discussion thus far has indicated that the linear sequence of bases on the DNA molecule is in some way related to the type of proteins synthesized by the cell. It is now known that this sequence is, in fact, a *genetic code* that can be translated into the amino acids of proteins. The purine and pyrimidine bases of the DNA molecule (adenine, thymine, cytosine, and guanine) are the "letters," or "alphabet," of the genetic code. All the genetic information contained in DNA can thus be figuratively represented by four letters: A, T, C, and G. Groups of three bases in linear sequence, which are usually called *base triplets* or *codons,* constitute the "code words" that are translated into one or another of the basic set of 20 amino acids found in proteins (see Table 7-1). An extraordinary feature of the genetic code is that these code words specify the same amino acids in every life form on this planet.

genetic code

base triplets
codons

Various combinations of the four bases in groups of three allow for 64 possible different codons ($4 \times 4 \times 4$). Since there are only 20 amino acids to be coded for, over 40 of these codons would seem to be redundant. In view of the fact that 61 of the 64 base triplets are known to code for amino acids, it must be concluded that a particular amino acid *may be specified by more than one codon;* that is, the genetic code is degenerate. Indeed, the property of *degeneracy* probably lessens the effects of "mistakes" (mutations) in the genetic code. The remaining three codons that do not specify amino acids are chain termination signals, or stop signals (see below).

degeneracy

Although the genetic program for proteins is contained in the base sequences on the DNA template, DNA itself does not synthesize proteins. The translation of the code into amino acids is accomplished by mRNA. Recall that mRNA is synthesized by transcription from (on) certain portions of DNA that contain the genetic code. The sequence of complementary bases on the mRNA molecule can thus be considered a working copy of the genetic code, or at least the complement of the code in which the RNA base uracil (U) has been substituted for the DNA base thymine (T).

Some examples of the mRNA codons and the amino acids they code for are shown below. (A list of standard abbreviations of amino acids is given in Chapter 7.)

Codon	*Amino acid*	*Codon*	*Amino acid*
UUU, UUC	Phe	AGU, AGC, UCU	Ser
UAU, UAC	Tyr	CGU, AGG, AGG	Arg
UUA, CUU, CUC, CUA	Leu	GGU, GGC, GGA	Gly
GUU, GUC, GUA, GUG	Val	ACA, ACG, ACC, ACU	Thr
AUG	Met	CAU, CAC	His

Chain termination signals: UAA, UGA, UAG

PROTEIN SYNTHESIS: TRANSLATION OF THE GENETIC CODE

We have seen that proteins are macromolecules consisting of one or more polypeptide chains. The chains, in turn, are composed of specific sequences of amino acids, linked together by peptide bonds. It should be evident by now that the polypeptide chains of proteins are not assembled at random or, in other words, in a hit-or-miss fashion. On the contrary, the sequence of each amino acid residue is *genetically specified*. Moreover, it is the differences in the sequences that determine the differences in the types of protein molecules synthesized by cells—and proteins ultimately determine the structure and function of living organisms. The enormous number of proteins that can be formed by possible combinations of the 20 amino acids accounts entirely for the great variety of different life forms on earth. In fact, the only real difference between an elephant and a flea is that each is composed of different proteins, which account for all structural and functional differences between the two organisms. Living systems synthesize protein molecules, amino acid unit by amino acid unit, according to the program of the genetic code embodied in their DNA. *Translation* is the process by which a specific sequence of base triplets—transcribed from DNA onto messenger RNA—is finally expressed as a specific sequence of amino acid residues in a protein molecule.

translation

Translation Mechanism

ribosomes

All three species of RNA participate in protein synthesis. The *ribosomes* provide the machinery on which the polypeptide chains are assembled. These organelles are made up of two subunits: a smaller subunit that holds the strand of mRNA and a larger one that binds tRNA molecules (see Fig. 11-7). In the cytoplasm of the cell, one or more ribosomes "thread into" the mRNA strand at the starting point and move along it sequentially. The process resembles a tape running through a playback machine. Usually, a group of ribosomes known as a *polyribosome*, or *polysome*, is associated with a single mRNA strand.

polysome

The codon sequence on the mRNA is not translated directly into amino acids. The actual translation of the code requires the presence of intermediates, that is, tRNA molecules, which deliver the appropriate amino acids to the mRNA strand on the ribosomes. The structure of the small tRNA molecules (see Fig. 11-8) indicates their function in the translation machinery. One end of the molecule binds a specific amino acid; the other end contains an exposed triplet of unpaired bases. The exposed base triplet, or *anticodon*, is complementary to the codon for the particular amino acid carried by that tRNA molecule. In other words, a specific codon on the mRNA strand will be "recognized" by the complementary anticodon of a tRNA molecule, not by the amino acid itself. The attachment of appropriate amino acids to tRNA carriers requires an investment of energy (ATP) and the catalyzing action of specific aminoacyl-tRNA synthetases, or *activating enzymes*. There is at least one of these enzymes for each amino acid. The carboxyl group of the amino acid is first energized by ATP

anticodon

activating enzymes

$$\text{amino acid} + \text{ATP} \rightarrow \text{aminoacyl-AMP} + \text{PP}_i$$

and the activated amino acid is then transferred to the tRNA molecule.

$$\text{aminoacyl-AMP} + \text{tRNA} \rightarrow \text{aminoacyl-tRNA} + \text{AMP}$$

aminoacyl-tRNA

The formation of *aminoacyl-tRNA complexes* is not only necessary for the transport of amino acids to ribosomes. Protein synthesis, that is, the formation of peptide bonds between individual amino acid units, is uphill work; it requires an investment of energy. Some of the energy of the two high-energy phosphate bonds expended in the above reactions is used to drive the incorporation of these amino acids into the polypeptide chains that are being synthesized on the ribosomes.

The key to the translation process is the temporary pairing (by hydrogen bond formation) between the anticodon triplet of a specific aminoacyl-tRNA and a complementary triplet on the mRNA strand. By this means, the appropriate amino acid, corresponding to the mRNA codon, is inserted in its proper sequence on the polypeptide chain.

The translation mechanism (Fig. 12-4) proceeds as follows:

1. Translation begins at an initiation codon* on the mRNA strand. The ribosomes that are involved in the synthesis of the polypeptide chain thread into the mRNA and move along the strand, codon by codon.
2. Two tRNA molecules, each loaded with a specific amino acid, briefly occupy two binding sites on the larger ribosomal subunit. Complementary base pairing takes place between the tRNA anticodons and the codons on the segment of the mRNA strand held temporarily by a binding site on the smaller subunit.
3. At each stage, enzymes on the ribosome catalyze the formation of a peptide bond between the amino acid carried by the first tRNA and the amino acid carried by the next.
4. The ribosome then moves up one triplet farther on the mRNA strand, displacing the first tRNA, which detaches and is free to pick up another amino acid.
5. Additional tRNA's, carrying amino acids, approach to take up their positions on complementary codons in the mRNA sequence.
6. As the ribosomes move along, the polypeptide chains forming on their surfaces are increased each time by one amino acid unit. Thus, ribosomes further along the messenger strand have longer polypeptide chains than those at the be-

* Part of the initiation signal, or start signal, for protein synthesis is the codon AUG, which also codes for the amino acid methionine. The initiation process appears to be a rather complex one, possibly involving the attachment of a modified aminoacyl-tRNA at this codon.

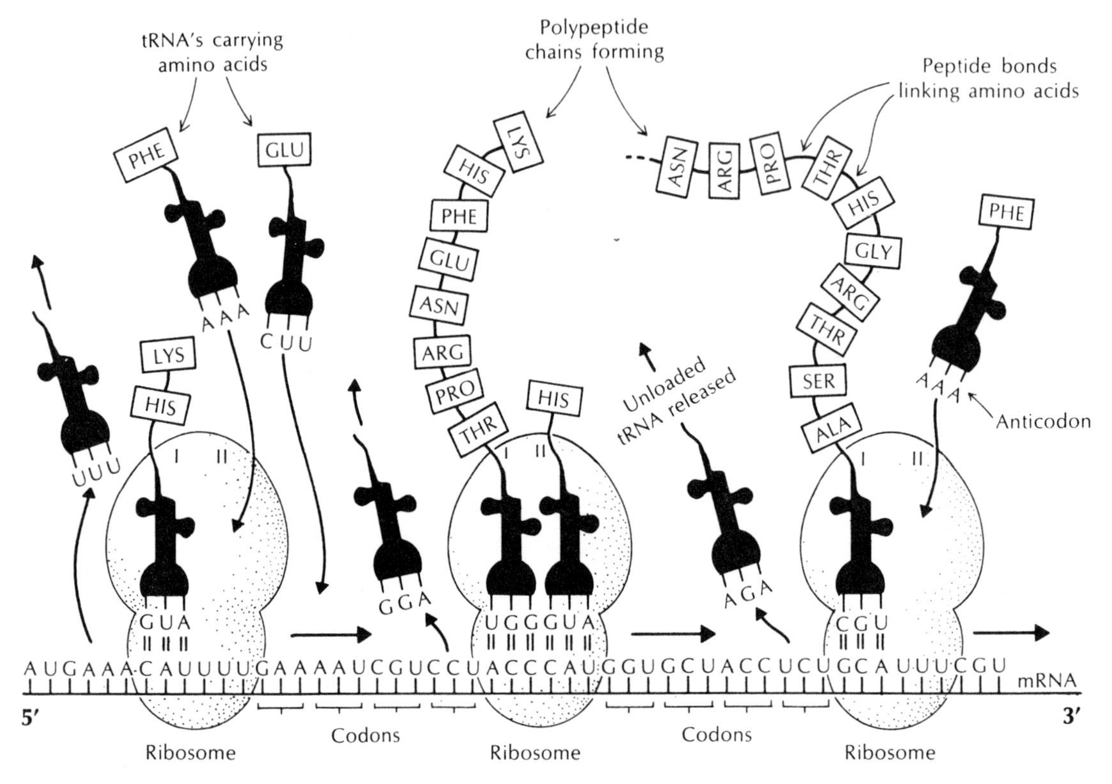

Fig. 12-4. Translation on a polyribosome. A group of ribosomes threaded onto a strand of mRNA are moving along its length, "reading" the code words, or codons. At each codon, a tRNA with the appropriate amino acid is temporarily bound. Each polypeptide chain (one on each ribosome) steadily grows in length until its ribosome reaches the chain termination signal. The completed chain is then released into the cytoplasm. It should be noted that mRNA is translated in the 5′ → 3′ direction.

ginning of the strand. When the ribosome reaches the stop signal on the mRNA, it discharges the completed polypeptide into the cytoplasm and threads on again at the start to reengage in protein synthesis. As noted previously, the *chain termination signals,* or stop signals, on mRNA are UAA, UGA, or UAG. There are no tRNA's with complementary anticodons for these codons. Instead, they are "recognized" by certain cell proteins *(release factors),* which bind to them and facilitate the release of the final polypeptide product from the terminal ribosome.

chain termination

release factors

Posttranslational Modification

Many newly-synthesized proteins subsequently undergo additional chemical alterations in the cell to enable them to fulfill specific structural or functional roles. This process is called *posttranslational modification*. For example, some proteins have carbohydrate groups attached to them (by *glycolsylation*), making them glycoproteins. Other protein products of the cell are synthesized in the form of much larger precursor molecules that are converted to active molecules only when the excess amino acid residues they contain are removed by proteolytic enzymes. The hormone, insulin, is a good model for this type of posttranslational modification. The mature insulin molecule is a small polypeptide containing a total of 51 amino acid residues. It is synthesized in islet cells of the pancreas as a much longer precursor, called *preproinsulin* (about 100 amino residues long), then trimmed down to a second precursor, *proinsulin* (about 80 residues long), and finally converted to its active smaller version.

posttranslational modification

glycosylation

DNA; RNA; Protein: Which Came First?

Our present knowledge of the genetic code is the outcome of intensive investigation, aided by the development of innovative techniques. The beautiful simplicity of the code and the basic modes of replication, transcription and translation have survived billions of years of evolutionary change. We now know that the genetic code, and the mechanisms associated with it, are *universal,* that is, they are substantially the same for all living systems on this planet. The flow of genetic information from DNA to RNA to protein (Fig. 12-5) is known as the central dogma. However, with regard to the origins of life on earth, there is a classic chicken-and-egg paradox here, namely, which came first, DNA or proteins? In today's living organisms, proteins cannot exist without DNA because only DNA contains the genetic blueprint for making them. At the same time, DNA cannot exist without proteins because protein enzymes perform most of the chemical work of replication, transcription, translation, DNA self-repair, etc. So which came first? A plausible solution to this riddle emerged with the discovery of ribozymes (RNA enzymes, see also page 140). RNA was once considered to be a sort of passive middleman in the transfer of genetic information from DNA to proteins. The existence of RNA enzymes led biologists to the view that life may well have started in an 'RNA world,' in which neither DNA nor

universal code

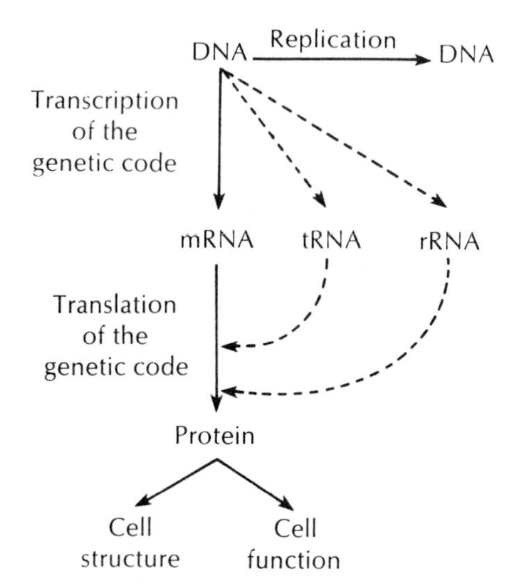

Fig. 12-5. Flow of genetic information in the cell.

proteins could have come first because both are descendants of RNA. In this pre-DNA, pre-protein domain, the earliest living organisms had self-replicating RNA molecules, and catalyzed all their chemical reactions with RNA enzymes. With the passage of time, these early RNA molecules learned to synthesize proteins, which proved to be faster and more versatile enzymes, and DNA, a more specialized repository of genetic information. RNA then assumed the more limited roles it plays in the modern cell, except that it still retains some of its original catalytic functions. But no one has yet figured out how RNA molecules, which are rather complicated structures, managed to get made in the first place from the raw materials in the non-living primordial soup.

GENES

genome

The size of the *genome,* the total amount of DNA in a representative cell of any given organism, generally increases with the degree of evolutionary complexity of the organism. The human genome, for example, has a total of about three billion base pairs arrayed on 46 chromosomes. How many of these base sequences are *genes?* And how do we define the gene? The best definition of a gene is in terms of modern molecular concepts: it is a sequence of DNA bases that codes for a polypeptide (protein), or an RNA product. In these terms, there are essentially two classes

genes

of genes, *protein-coding genes* that are transcribed to messenger RNA, and genes that do not code for proteins, but are used instead for the transcription (synthesis) of ribosomal RNA and transfer RNA, i.e., *rRNA genes* and *tRNA genes*.

protein-coding genes

rRNA genes

tRNA genes

The human genome is estimated to contain about 100,000 genes, which accounts for only a small percentage of its total DNA content. Some of the excess is due to the presence of repetitive gene and non-gene sequences. Human DNA, for example, contains many multiple copies of protein-coding and RNA-coding genes. Additionally, as we noted above, most genes in higher organisms are discontinuous, or *split genes*. The coding portions of the genes *(exons)* are interrupted by quite lengthy non-coding portions *(introns)*. Other findings indicate that clusters of related genes on a chromosome are often interrupted by non-transcribed base sequences, called *spacers*. In most cases, the spacer segments are longer than the genes. Spacers should not be confused with introns; they are found between genes, rather than within them. The most prevalent repeated non-gene sequence in mammalian genomes is the *Alu* sequence, a distinctive chain of up to 300 base pairs located in introns and spacers. About half a million copies of the *Alu* sequence have been identified in the human genome, and it accounts for over 5% of the total content of DNA in the genome. The role of this 'silent,' or what is sometimes called 'junk' DNA in the genomes of humans and other higher organisms is largely unknown at the present time.

split genes

exons

introns

spacers

Gene Regulation: The Operon Model

It should be apparent that although all cells in the body are descended from one cell (the fertilized ovum) and all cells therefore contain identical DNA, they do not all produce identical proteins, nor are they all structurally and functionally identical; that is, *individual cell types translate only a part of the total genetic information that is available to them*. The process by which the cells of a developing embryo express different genes and assume different morphological (*morpho,* form or shape) and functional characteristics is called *differentiation*. In terms of molecular biology, how can we account for the fact that a liver cell and a neuron, for example, are structurally and functionally distinct? Obviously, cells must have mechanisms that control which of their genes will be expressed, and which suppressed. These mechanisms are by no means fully defined or understood at the present time, but valuable insights into some of the on-off switches that control gene expression and protein synthesis have been provided by Jacob and

differentiation

Monod's* study of the lactose operon system in *Escherichia coli,* a strain of bacteria that normally inhabits the human intestine. These microorganisms exhibit a dynamic pattern of response to their environment that is now considered to be characteristic of all living systems. When they grow on a medium containing the sugar lactose, they synthesize three enzymes specific for the metabolism of this substrate; when the culture medium contains little or no lactose, the three genes responsible for the synthesis of the enzymes appear to be switched off. This fine adjustment is mediated by the coordinate function of a genetic unit called an *operon*.

operon

An operon consists of:

structural genes

1. A group of adjacent *structural genes* on a DNA strand, transcribing a single mRNA molecule. These genes dictate the amino acid sequence of a set of functionally related proteins. (In the lactose operon, three structural genes code for the three enzymes that are required for the utilization of lactose.)

operator

repressor

2. An *operator site*—a small section of DNA adjacent to the structural genes, made up of a short sequence of bases that fits into the binding site of a specific *repressor* protein.

promoter

3. A *promoter site*—an additional section of DNA, adjacent to the operator, that attaches to the enzyme RNA polymerase, thus initiating transcription of mRNA from the structural genes.

regulatory gene

The function of the operon is controlled by a *regulatory gene* that codes for the synthesis of a specific repressor protein. By means of these repressors, regulatory genes control the transcription of structural genes and thus, indirectly, the rate of protein synthesis by the cell. *When the repressor binds to the operator site, it blocks the promoter and switches off the transcription of the structural genes*. Repressor proteins provide a versatile system of on-off switches, since they can be functionally modified in a number of ways (Fig. 12-6). In the lactose operon, the repressor is *inactivated* when it combines with an *inducer,* which, in this instance, is lactose. The inactivated repressor cannot bind the operator, and the transcription of the three structural genes that code for the necessary enzymes can then proceed. This is a classical example of *enzyme induction* (described in Chapter 8).

inducer

enzyme induction

In other operon systems, the regulatory gene may code for an

---

* French geneticists who won the 1965 Nobel Prize for elucidating replication and translation mechanisms.

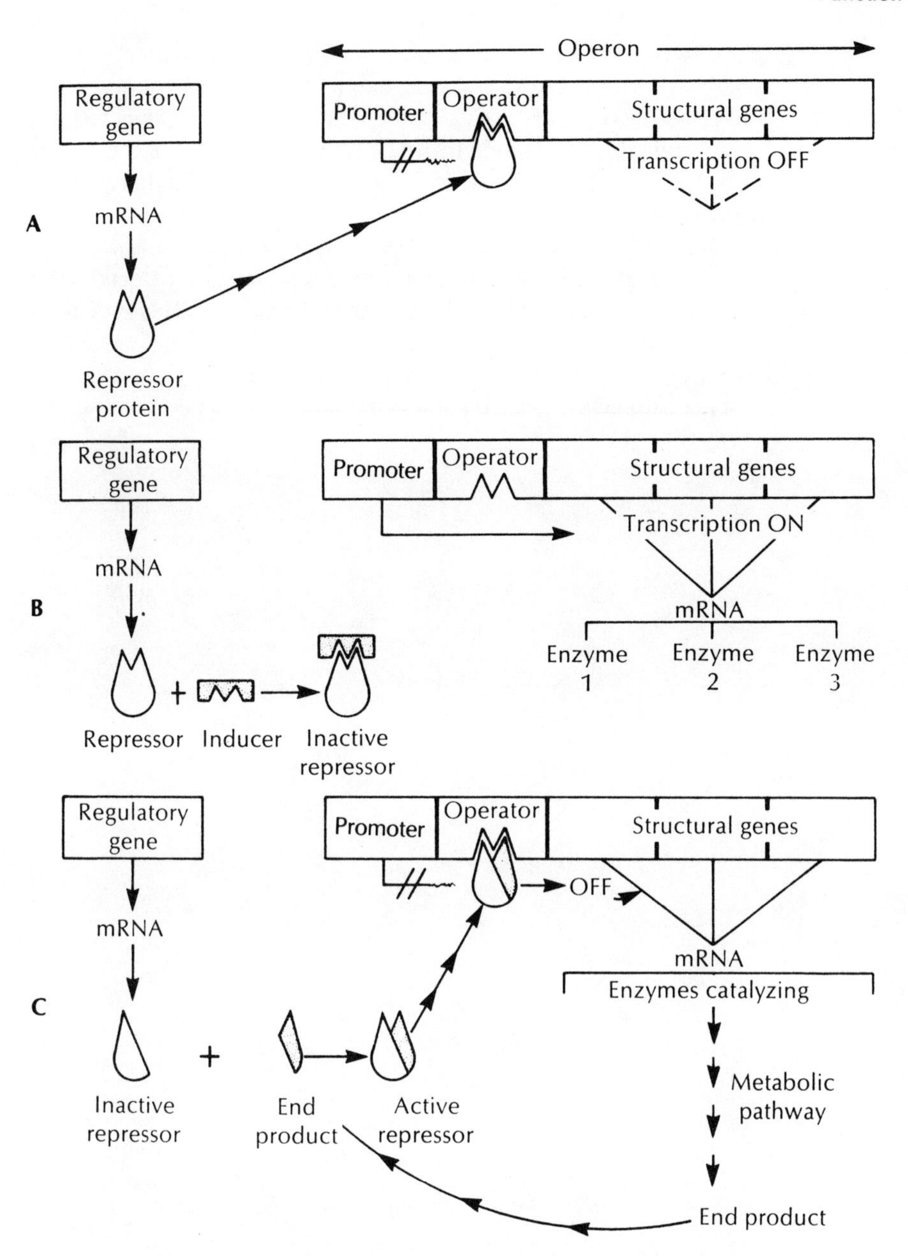

Fig. 12-6. Models of operon function. **A,** Regulatory repressor binds to the operator site and blocks transcription of the structural genes. **B,** Enzyme induction: the repressor is inactivated by an inducer substance (usually a substrate), permitting transcription of structural genes. **C,** Enzyme repression: an inactive repressor produced by the regulatory gene is activated by a metabolic end-product and blocks transcription of the structural genes.

enzyme repression

inactive precursor of the repressor, which is *activated* when it combines with the accumulating end product of a particular metabolic pathway. The activated repressor then switches off the structural genes that code for the enzymes involved in the chemical reactions of that pathway. This process is called *enzyme repression*.

The coordinated activity of the genes of a cell thus enable it to adjust metabolically to changes in its internal and external environment. Furthermore, feedback-control models of this type point out possible mechanisms by which cells of multicellular organisms may differentiate from each other. The dynamics of operon systems reemphasize the basic fact that genes are, in essence, a program for survival.

13

MOLECULAR GENETICS

DEFECTS IN THE CODE—GENETIC DISEASES

Knowledge of how the genetic code controls protein synthesis has helped to clarify the nature of many inborn metabolic diseases. Inheritance in man is determined by the information carried on 46 chromosomes (22 pairs of autosomes and 1 pair of sex chromosomes), which have been estimated to contain as much as a million genes. The genes are paired; that is, one *(allele)* of each pair is derived from one parent. An allele may be characterized as dominant or recessive; the dominant one blocks to a varying extent the expression of the recessive.

allele

As noted in Chapter 12, genes dictate the ultimate structure and function of an organism by means of a code that specifies the types of protein molecules it will produce. The message of the code is expressed in the sequence of amino acid residues in each protein molecule; this sequence, in turn, endows the protein with special properties that enable it to play its role in the living system. An "error" in a gene, or in the translation process, could thus have widespread repercussions in the organism, depending on the relative importance of the product of that gene.

Changes in genes are called *mutations;* they are produced by changes in the sequence of bases in the gene. The effects of such changes in base sequences can best be illustrated by the ordinary typographical errors that might occur in the spelling of a word:

mutations

Correct code word (base sequence): Paper
Deletion error: Aper
Insertion error: Pamper
Substitution error: Piper
Scrambling error (nonsense): Eprpa

Note that the errors completely change the *meaning;* that is, the amino acid that is specified by the correct sequence of bases.

Most gene mutations are apparently the result of random accidents occurring during the replication process; others are known to be caused by certain chemical agents *(mutagens)* and ionizing radiations such as x-rays and cosmic rays. The mutations of a gene in a somatic cell is self-limiting, since the change will be exhibited only by the descendants of that particular cell. However, a mutation in the DNA of a *germ cell,* that is, a spermatozoon or ovum, can be transmitted from parent to offspring, from one generation to the next.

mutagens

germ cell

Metabolic Defects Caused By "Missing Enzymes"

A mutation may result either in defective function (block) or in the absence of a gene coding for the synthesis of a particular enzyme. Conditions of this type are generally called *inborn errors of metabolism* (see also Chapter 8). The mechanism whereby this leads to a metabolic disease can be outlined as follows:

inborn errors of metabolism

1. A normal metabolic pathway for the conversion of substance A to substance D involves a sequence of three chemical reactions, each catalyzed by a specific enzyme:

$$\begin{array}{ccccccc} & \text{Gene I} & & \text{Gene II} & & \text{Gene III} & \\ & \downarrow & & \downarrow & & \downarrow & \\ & \text{Enzyme I} & & \text{Enzyme II} & & \text{Enzyme III} & \\ & \downarrow & & \downarrow & & \downarrow & \\ \text{A} & \longrightarrow & \text{B} & \longrightarrow & \text{C} & \longrightarrow & \text{D} \end{array}$$

2. If Enzyme III is defective or absent because Gene III is mutated, the reaction sequence C → D will not take place.
3. Metabolic disease may then result from:
   a. Lack of substance D
   b. Toxic accumulation of the intermediary substance C
   c. Toxic accumulation of substances A and B if the reaction sequences are reversible:

$$A \rightleftarrows B \rightleftarrows C \xrightarrow{\text{Block}} D$$

d. Abnormal production of another substance, for example, substance X, by an alternative metabolic pathway:

$$A \rightarrow B \rightarrow C \xrightarrow{\text{Block}} D$$
$$\searrow X$$

Events of this type cause a spectrum of biochemical disorders. The best example is a disease called *phenylketonuria* (PKU), which is characterized by marked mental retardation in children. PKU is transmitted by an autosomal (non-sex-linked) recessive gene; that is, the diseased offspring can inherit a pair of identical defective PKU alleles from normal heterozygous parents. The missing enzyme in PKU is *phenylalanine hydroxylase,* a liver enzyme that catalyzes the conversion of the amino acid *phenylalanine* to another amino acid, tyrosine. The block of this metabolic pathway causes accumulation of phenylalanine in the blood and excessive production of the substance phenylpyruvic acid by an alternative pathway. The latter is excreted in the urine. The excess of phenylalanine and phenylpyruvic acid in the body tissues and body fluids is responsible for the symptoms of the disease.

PKU

phenylalanine

Genetic Diseases Caused By Point Mutations

A *point mutation* occurs when a single purine or pyrimidine base in the sequence of nucleotides of a gene is replaced by a different base. This change in one codon can result in the insertion of a different amino acid into the polypeptide under the control of that gene.

point mutation

An important example of an altered protein produced by a point mutation is the abnormal hemoglobin *HbS* found in *sickle cell anemia*. The disease is essentially a severe hemolytic anemia caused by excessive loss of red blood cells. The abnormal cells have a life span of 60 days, as compared with the 120-day norm. Sickle cell anemia takes its name from the bizarre sickle shapes assumed by abnormal red blood cells (those containing abnormal HbS), particularly in blood with a low oxygen tension.

HbS

sickle cell anemia

The point mutation is expressed in the globin (protein) portion of the *hemoglobin* molecule. Hemoglobin (molecular weight about 68,000 d) is made up of four polypeptide subunits bound to four iron-containing complexes called heme. The molecule is symmetrical and composed of identical halves. Each pair of polypeptides consists of some 290 amino acid residues. In normal adult hemoglobin, *HbA,* the sixth amino acid on the β-polypeptide chain is *glutamic acid*. However, in the sickle cell hemoglobin, HbS, the

hemoglobin

HbA

glutamic acid position is occupied by *valine*. Thus, the only demonstrable difference between normal HbA and defective HbS is the substitution of one valine for one glutamic acid, a change in 1 out of a total of 290 amino acid residues. This seemingly insignificant change distorts the normal conformation of the hemoglobin molecule and accounts for the aggregation of HbS molecules at low oxygen tensions.

sickle cell trait

The disease is transmitted as an autosomal (non–sex-linked) recessive. Persons suffering from *sickle cell anemia* inherit a pair of identical defective recessive alleles from their parents. *Sickle cell trait* is exhibited in clinically healthy, heterozygous carriers who have inherited one normal dominant gene and one abnormal recessive gene. In this case, the dominant gene does not altogether cancel out the expression of the recessive gene; the red blood cells of carriers of the *trait* contain a mixture of both normal HbA and defective HbS.

MODERN TRENDS IN MOLECULAR GENETICS

The past decade has witnessed the introduction of increasingly sophisticated tools for investigating gene structure and function—which has led, in turn, to major advances in our knowledge of molecular genetics. In the remaining sections of this chapter, we shall discuss a few of the more important aspects of this large and expanding branch of science.

Viruses

viruses

Viruses are extremely small packets of genetic information made up of an outer shell of protein and an inner core of either DNA *or* RNA. Unlike cellular forms of life, they do not possess both varieties of nucleic acid. It is difficult to classify viruses in terms of standard criteria for 'living' and 'non-living' matter. They have been most appropriately described as "packets of infectious genetic material" and "genes looking for a place to function." Viruses lack the biosynthetic and metabolic apparatus of the cell. They are therefore totally dependent on the cellular machinery and chemical resources of the *host cells* they invade. The infective portion of the chemically mature virus particle, or *virion,* is the nucleic acid; the protein coat protects the nucleic acid core from enzymatic attack, and also appears to play a role in the penetration of host cell membranes. When the virus particle enters a cell, it preempts the biosynthetic machinery of the cell to replicate itself.

host cells
virion

Viruses are responsible for a large number of diseases in plants and animals. In man, they are known to be the causative agents of such diseases as rabies, smallpox, poliomyelitis, influenza, measles, mumps, chickenpox, yellow fever, and AIDS.

Scientists have found viruses very useful instruments for the study of genetic molecules and genetic mechanisms. Furthermore, insofar as the health of humans, other animals, *and* plants is concerned, viruses have been a constant challenge to investigators to devise more effective ways of preventing and treating virally-caused diseases.

Many studies on viral reproduction in cells have been done on *bacteriophages,* DNA viruses that infect bacterial cells (Fig. 13-1). Phages act in one of two possible ways after invading a bacterium:

bacteriophages

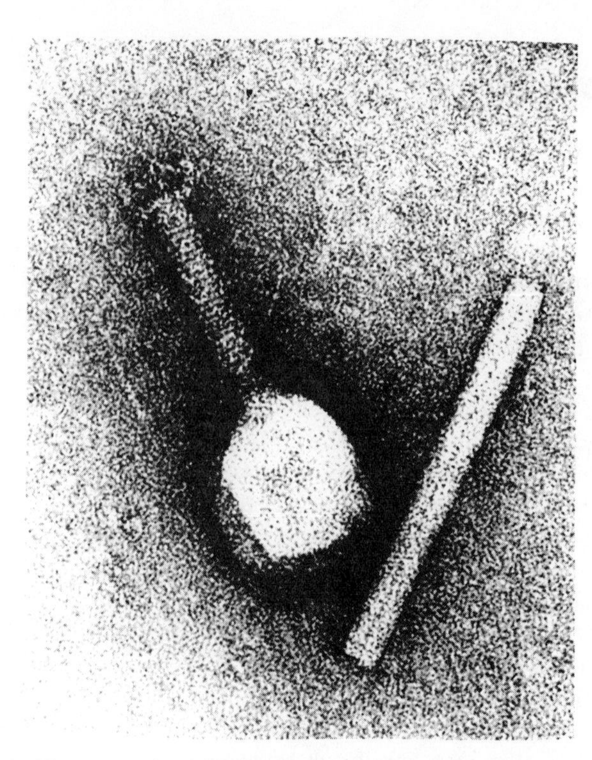

Fig. 13-1. Electron micrograph of a DNA and RNA virus. The flask-shaped DNA virus is a T4 bacteriophage; the rod-shaped RNA virus is a tobacco mosaic virus. (×450,000.) (Courtesy F. A. Eiserling. From Brooks, S. M.: Basic biology; a first course, St. Louis, 1972, The C. V. Mosby Co.)

1. The phage particle may monopolize the cellular machinery and actively replicate new viral DNA molecules and synthesize new viral protein. When this occurs, the DNA and RNA of the host cell cease to function and the synthesis of host cell proteins stops abruptly. This process ends with the destruction of the host bacterium by *lysis* (bursting open) and the liberation of 100 or more new virions.
2. The DNA molecule of the invading phage virus may become incorporated into the bacterial chromosome and persist there in a rather dormant state, replicating only at each division of the bacterial cell.

lysis

Acute viral infections in man generally parallel the first process described above. Viral infection is usually curbed by an immunologic reaction against the viral antigen. However, an increasing amount of attention is now being focused on the second, dormant-type, invasive process.

Oncogenic Viruses

oncogenic viruses

transformation

The existence of *oncogenic viruses* (viruses that cause cancer) has been known for many years. The characteristic property associated with these viruses is that they can produce malignant changes, termed *transformation,* in the cells they invade. The most striking characteristic of malignantly transformed cells is their ability to proliferate continuously in an uncontrolled manner. This is associated with a tendency to invade surrounding normal tissues, and to spread (metastasize) to distant sites in the body. Malignant transformation of cells occurs when the viral genome is integrated or stitched into the DNA of the host cell, after which it becomes a part of the host's genome. Some oncogenic viruses are DNA viruses, and their insertion into the cell's DNA is obviously feasible. RNA viruses, on the other hand, have to use different strategies to transform cells. In general, the behavior of RNA viruses is reminiscent of the primeval 'RNA world' described in the preceding chapter. For example, using the host cell's facilities, various species of RNA viruses can not only transcribe themselves (i.e., synthesize more viral RNA, using themselves as templates), they can transcribe DNA. The latter class of RNA viruses are known as *retroviruses.* When a retrovirus invades a cell, it uses a viral enzyme, called *reverse transcriptase,* to synthesize a complementary double-stranded molecule of DNA. This DNA transcript of the retroviral RNA can then incorporate itself into the host's genome. The human immunodefi-

retroviruses
reverse transcriptase

ciency virus (HIV) which causes AIDS is a notable example of a retrovirus.

It was soon demonstrated that transformed cells were, in many cases, expressing the incorporated genes of the oncogenic viruses that had originally invaded them. These viral genes subsequently came to be known as *oncogenes*. Essentially, oncogenes are changed versions of normal cellular genes (or *proto-oncogenes*) whose normal protein products control cell growth and cell reproduction. Of interest here is the finding that the genes of the cancer-causing viruses are not really alien genes at all; they are actually somewhat altered copies of cellular genes that were previously 'captured' by the viruses from the genomes of cells they had once infected.

oncogenes
proto-oncogenes

Currently, some 20% of all cancers are believed to be virus-induced. For example, three DNA viruses have been linked to common cancers: some strains of human papilloma virus to cancer of the cervix of the uterus, Epstein-Barr virus to lymphoma and cancers of the nose and throat, and hepatitis B virus to liver cancer. An RNA retrovirus (HTLV) is known to cause adult T cell leukemia in humans, and people infected with the AIDS retrovirus have a substantially increased risk of developing certain tumors, such as lymphomas and Kaposi sarcomas.

At this point, it is important to note that other factors besides viruses can convert normal cellular genes to oncogenes. These include DNA-mutating agents, such as radiation and carcinogenic chemicals. Current concepts of gene-cancer links are summarized in Fig. 13-2.

Genetic Engineering: Recombinant DNA Technology

Our insights into how the genomes of all biological systems, from viruses to man, are organized, expressed, and controlled have greatly expanded with the use of two powerful research tools that were developed during the 1970's, namely, rapid methods for determining nucleotide sequences in DNA, and the recombination and cloning of DNA. These techniques enabled scientists to insert DNA segments, from any source, into the DNA of bacteria and other cell types, including mammalian cells maintained in tissue culture. The DNA product of such manipulation is called *recombinant DNA;* the technique is popularly known as gene-splicing.

recombinant DNA

The insertion of sections of DNA into a DNA molecule became a reality with the discovery and isolation of a group of remarkable bacterial enzymes, known as *restriction endonucleases,* that have the ability to cleave DNA molecules at certain sites with highly

restriction endonucleases

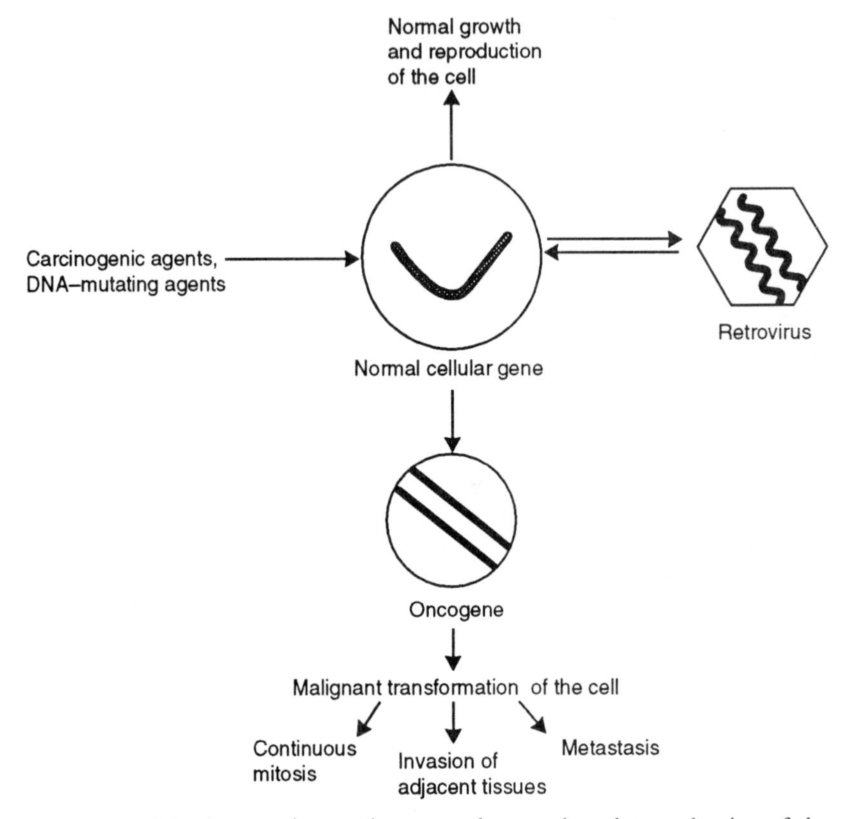

Fig. 13-2. A normal gene that controls growth and reproduction of the cell can be altered by DNA-mutating agents, or picked up by a retrovirus, altered slightly, and re-inserted in another normal cell. In either case, an oncogene is formed in the target cell, and it is transformed to a malignant cell.

specific base sequences. Such enzymes are synthesized by many microorganisms; they apparently evolved to attack the DNA of the invading viruses. When a DNA molecule is cleaved by a restriction enzyme, the cut ends are "sticky"; and the molecule will recombine, by complementary base pairing, with another DNA segment that has been cleaved by the same enzyme. If the inserted segment is from a DNA molecule of a different organism, a recombinant "hybrid" DNA molecule will be produced. The two DNA strands in the recombinant molecule are then spliced together by the enzyme *DNA ligase*. The inserted DNA segment in a recombinant molecule is often called the *passenger;* the rest of the DNA in the molecule is called the *vehicle*. Once the passenger is sealed to the vehicle, the two components generally replicate together and transcribe their relevant mRNA's.

DNA ligase
passenger
vehicle

The most convenient vehicles for recombinant DNA experiments are tiny bodies composed of circular double-stranded DNA, found in some bacteria. These structures, called *plasmids,* exist apart from the bacterial chromosome.* However, like the bacterial chromosome, plasmids are composed of gene sequences, replicate themselves, and transcribe mRNA. By the combination of various techniques, different DNA passengers can be spliced into plasmids; and the hybrid plasmids can be introduced into bacterial host cells (Fig. 13-3). Bacteria reproduce asexually with great rapidity. Given the appropriate culture media, bacteria bearing hybrid plasmids will quickly multiply to form a colony.

plasmids

Since the DNA passengers faithfully reproduce copies of themselves at each division of the bacterial cell, it is customary to call such experiments *DNA cloning* (or molecular cloning). The term *clone* is usually used (either as a noun or a verb) to indicate a line of cells propagated by culturing a single cell. In the context of recombinant DNA experiments, the term *clone* has two applications: it can refer to a line of transformed cells that are all descended from the original cell bearing the recombinant DNA molecule; or in the more limited sense, it refers specifically to the spliced fragments of foreign DNA that are being propagated in the transformed host cells and in their descendants. The advantages of DNA cloning are immediately evident; at the will of the investigator, individual genes (or specific sequences of DNA), from *any* source, can be inserted into an appropriate vehicle and will then replicate copies of themselves indefinitely.

DNA cloning

After years of research and testing, recombinant DNA technology has moved into the marketplace. From the beginning, it was recognized that besides greatly facilitating studies of the gene, the new technology had potentially valuable clinical, environmental, and agricultural applications. We now have a thriving genetic engineering industry creating new products at an accelerating pace. For example, recombinant DNA is now used for the manufacture of bulk quantities of pharmaceutical products, such as various drugs, vaccines, hormones, growth factors, clotting and anti-clotting factors, and so on. But modern recombinant biotechnology is also aimed at breeding superior plants and livestock, and curing genetic diseases.

---

* It should be noted that bacteria are *prokaryotes,* simple unicellular organisms that do not possess a true nucleus. The genetic apparatus of such organisms is distributed in the cytoplasm, in contrast to *eukaryotes,* organisms whose cells contain a distinct membrane-enclosed nucleus.

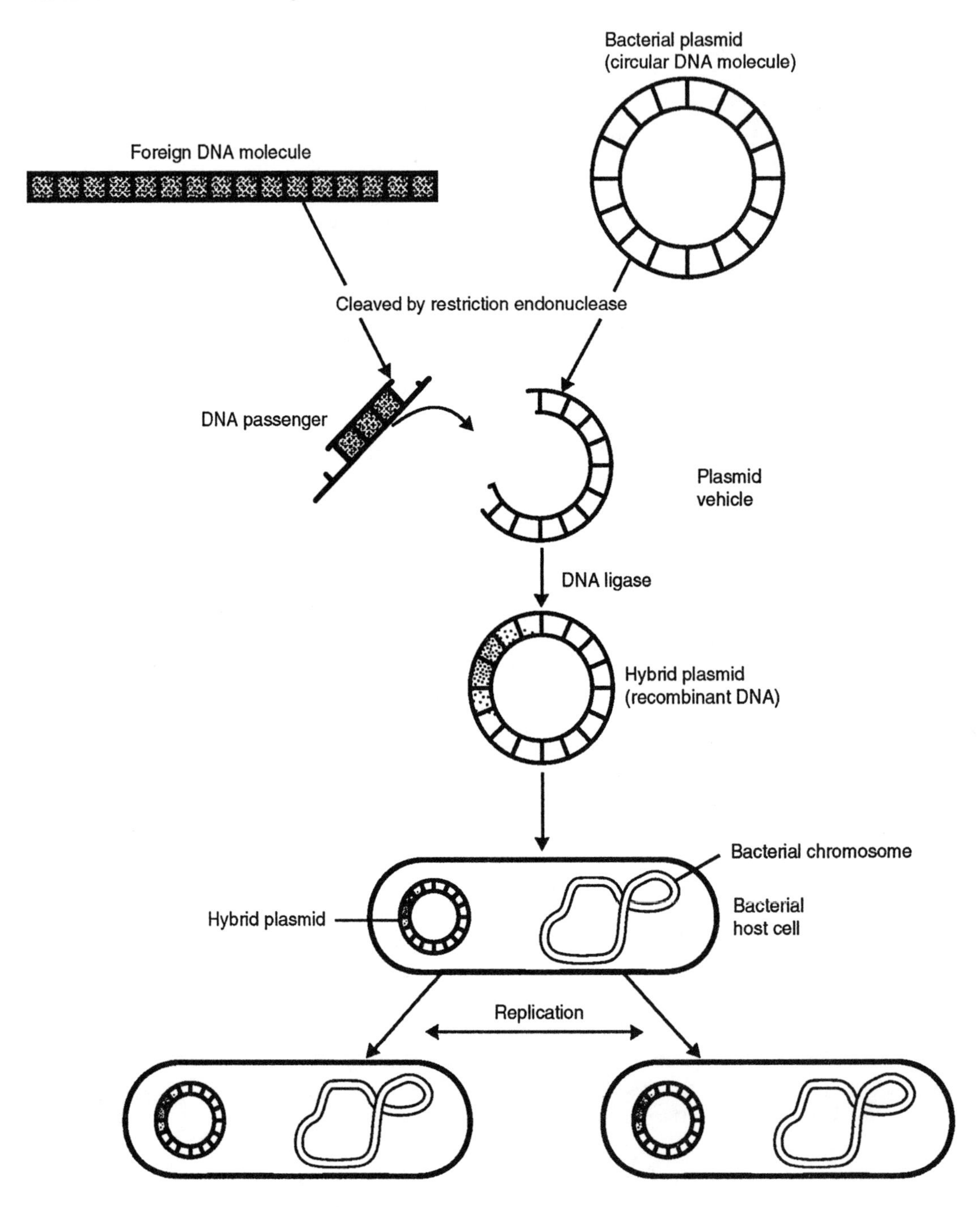

Fig. 13-3. DNA cloning: a segment of foreign DNA is spliced into a plasmid and introduced into a bacterial cell, where it replicates itself.

Gene Therapy

gene therapy

One clinical application of recombinant DNA technology that is being explored at the present time is *gene therapy*—the treatment of certain human genetic diseases by replacing a defective mutant gene with a normal one. Several rare but serious, and often fatal, genetic diseases are now targeted as candidates for gene therapy. Various techniques have been devised for cloning normal genes and inserting them into the DNA of appropriate target cells. However, problems involving gene transfer, inactivity of the transferred genes, and the transplantation of cells into patients have proved to be major technical obstacles that must be overcome before gene-based therapy can be successful. It has been rightly said that gene therapy will not reach its full potential until the replacement genes can be coaxed to work throughout the lifetime of the patient.

INDEX

Page numbers in **boldface** refer to figures. A page number followed by t indicates a table; a page number followed by n indicates a footnote.

F

Q

R

W

X

Z

Name____ Date____________________

CHAPTER 1

Key Terms

matter
element
atom
proton
electron
neutron
atomic mass unit (amu)
dalton (d)
atomic number
mass number
atomic weight
isotope
Periodic Law
metals
nonmetals
metalloids
halogens
inert (noble) gases
radioactivity

Study Questions

1. The halogens in Group VIIA of the periodic table all have _______ electrons in their outer energy levels.
2. The four most abundant elements present in human tissues are _______.
3. Elements in Group VIIIA at the far right of the periodic table are called the _______. They have little or no tendency to combine with other elements because _______.
4. What is the chemical symbol for each of the following elements: (a) mercury (b) potassium (c) sodium (d) copper (e) iron
5. Deuterium and tritium are isotopes of what element?
6. Given the chemical notation for a radioactive isotope of cobalt, $^{60}_{27}Co$, how many protons are present in its nucleus?
7. Convert: (a) 50°C to °F (b) 50°F to °C (c) 250 μm to mm (d) 1.35 L to cc
8. The two lightest elements are _______ and _______.
9. Which is the larger number: 10^{-10} or 10^{-2}?
10. Change the following numbers to nonexponential form: (a) 6.39×10^{-5} (b) 0.0071×10^{4}
11. Which is heavier: an object weighing 1 microgram or an object weighing 1 nanogram?

Name________________________________ Date________________

CHAPTER 2

Key Terms

chemical compound
chemical bond
chemical formula
organic chemistry
valence electrons
octet rule
ion
cation
anion
ionic bond
law of constant proportions
chemical formula
law of conservation of mass/energy
covalent bond
polar covalent bond
oxidation/reduction
hydrogen bond
molecular weight
formula weight
mole (mol)
Avogadro's number

Study Questions

1. The type of chemical bonds in which atoms share electron pairs are called ______ bonds.
2. Why is it incorrect to balance the chemical equation:

   $$H_2 + O_2 \rightarrow H_2O$$

   in this manner:

   $$H_2 + O_2 \rightarrow H_2O_2$$

3. An example of an element that exists as a diatomic molecule is ______.
4. One mole of *any* chemical substance will contain the same number of units (atoms, ions or molecules). This number is called ______.
5. Iron combines with oxygen to form iron oxide (rust) as follows:

   $$Fe + O_2 \rightarrow Fe_2O_3$$

   Balance this equation.
6. The oxidation number of an atom in a neutral (uncombined) state is ______.
7. In the equation: $A + B \rightarrow X + Y$ — which would be the *reactants* and which the *products*?
8. Given the atomic weight of potassium (K) is 39, that of sulfur (S) is 32, and that of oxygen (O) is 16, what would be the formula weight of the compound, potassium sulfate (K_2SO_4)?
9. How many electron pairs do the two nitrogen atoms share in forming a molecule (N_2)?
10. An example of a common positively charged polyatomic ion is ______. (Indicate the chemical formula of the ion and its charge.)

Name________________________________ Date________________

CHAPTER 3

Key Terms

dipole
hydronium ion
hydroxyl ion
hydrophobic interactions
dissociation
ionization reactions
acids
bases
proton donors
proton acceptors
neutralization reactions
pH
electrolytes
nonelectrolytes

Study Questions

1. What type of bonds hold aggregates of water molecules together? Describe *two* physical properties of water that result from these bonds.
2. A substance that accepts protons is defined as a _______.
3. Using the chemical formulas, give *two* examples of polyprotic acids.
4. The concentration of hydrogen ions $[H^+]$ in a test solution is 1.0 M/Liter. What is the pH of this solution?
5. The pH of human body fluids is maintained homeostatically at approximately _______.
6. The pH of pure water is _______.
7. Acids react with bases to form _______ and _______. This type of chemical reaction is termed _______.
8. What is the difference between a *strong* acid and a *weak* acid?
9. The hydration of CO_2 in red blood cells is greatly speeded up by the enzyme _______.
10. The bicarbonate ion is the conjugate base of what acid? (Give the chemical formula of the acid.)
11. Define a *buffer*.
12. The ion product constant (K_w) of water equals _______ at 25°C.
13. What is the chemical formula for the hydronium ion?

Name____ Date____________________

CHAPTER 4

Key Terms

organic chemistry
carbonyl group
alcohols
amino group
asymmetrical carbon atom
hydrocarbons
alkanes
alkyl group
isomer
alkenes
alkynes
carboxyl group
ester
ether
cyclic compounds
pyrimidines
purines
polymers
biomolecules
macromolecules
residues
biosynthesis
hydrolysis
precursor

Study Questions

1. The simplest hydrocarbon compound in the alkyne series is ______. (Indicate the name *and* formula of the compound.)
2. An organic compound that has a carbonyl group (—C=O) at the end of a hydrocarbon chain is classified as a ______.
3. A single carbon atom can form how many covalent bonds?
4. How many benzene rings are there in the molecules of compounds such as anthracene and phenanthrene?
5. The presence of ______ bonds in a hydrocarbon chain compound indicates that it is unsaturated.
6. The molecular formula for the compound, benzene, is ______.
7. The decomposition of large molecules into smaller molecules by reaction with H_2O is termed ______.
8. The molecules of all organic acids contain one or more ______ groups.

Name________________________________ Date____________________

CHAPTER 5

Key Terms

carbohydrate
monosaccharide
disaccharide
polysaccharide
hexose
pentose
aldose
ketose
optical activity
enantiomers
phosphorylation
ascorbic acid
glycogen
cellulose
glycogenesis
glycogenolysis
gluconeogenesis
glycoproteins
glycolipids
cerebrosides
gangliosides
proteoglycans
glycosaminoglycans
oligosaccharides

Study Questions

1. The chemical name for milk sugar is ______.
2. The building block unit (monomer) of starch and glycogen is ______.
3. The term *glycogenolysis* refers to ______.
4. Three monosaccharides, all having the formula $C_6H_{12}O_6$, are ______, ______, and ______.
5. An abundant polysaccharide of plant cell walls, which humans cannot digest, is ______.
6. A water-soluble vitamin that may be classified as a sugar acid is ______.
7. Oligosaccharide-containing proteins, such as those found on cell membranes, are generally termed ______.
8. Three examples of disaccharides, that are commonly present in foods, are ______, ______, and ______.

Name________________________________ Date________________

CHAPTER 6

Key Terms

lipids
saturated fatty acids
unsaturated fatty acids
polyunsaturated fatty acids
triacylglycerols
lipolysis
endogenous fat
exogenous fat
lipase
lipogenesis
phospholipid
lipid storage disorders
turnover
lysosomal enzymes
amphipathic molecules
micelles
lipid bilayers
steroids
cholesterogenesis
emulsion
carotene
antioxidant
oxidative stress
prostaglandins
plasma lipoproteins

Study Questions

1. A fatty acid consists of a hydrocarbon chain with a terminal _______ group.
2. Fatty acid molecules that have one or more double-bonded carbon atoms are said to be _______.
3. _______ consist of three molecules of fatty acids esterified to one molecule of glycerol.
4. A fat-soluble vitamin that is synthesized by bacteria in the human intestine is _______.
5. The first and second steps in the hydroxylation (activation) of cholecalciferol (vitamin D_3) occur in what two organs?
6. The sex hormones and bile acids are chemical derivatives of _______.
7. The drug, aspirin, and related nonsteroidal anti-inflammatory drugs, block the synthesis of what class of lipid substances in the body?
8. Tay-Sachs disease is a hereditary disorder caused by the defective turnover of _______ in nerve cells.
9. Two examples of biologically-important amphipathic lipids are _______ and _______.
10. The most widely distributed steroid in the body is _______.
11. Amphipathic lipids tend to assume two types of arrangements in aqueous media, namely _______ and _______.
12. Fat digestion depends on the emulsifying action of _______ in the small intestine.
13. The largest lipoprotein particles in the blood stream are the _______.
14. High blood levels of which group of lipoproteins is now associated with an increased risk of heart attacks?

Name____ Date____________________

CHAPTER 7

Key Terms

proteins
polypeptides
alpha-amino acids
amino acid residues
D- and L-isomers
R group (side chain)
zwitterion
peptide bonds
subunit
α-helix
β-pleated sheet
primary structure
secondary structure
tertiary structure
quaternary structure
denaturation
apoprotein
conformation
ligand

Study Questions

1. What group in an amino acid distinguishes it from other amino acids?
2. The proteins of all living organisms on earth are assembled from the same group of ______ amino acids.
3. Amino acids in neutral solutions exist as dipolar ions or ______.
4. The formula for the amino group is ______.
5. Two amino acids with sulfur-containing side chains are ______ and ______.
6. The strong covalent bonds between individual amino acid residues in a protein molecule are called ______.
7. The primary structure of a protein molecule is determined by ______.
8. Two examples of secondary structures in protein molecules are ______ and ______.
9. The three-dimensional shape of a protein molecule is termed its ______.
10. Specific compounds that bind to functional proteins, fitting like keys into locks, are generally called ______.
11. Disorganization of the characteristic three-dimensional shape of a protein molecule is called ______.
12. What element is present in proteins that is usually not present in carbohydrates and lipids?

Name____ Date____________________

CHAPTER 8

Key Terms

activation energy
catalyst
enzyme
ribozyme
active site
substrate
end product
isozyme
zymogen
apoenzyme
holoenzyme
prosthetic group
coenzyme
allosteric (regulatory) enzyme
affinity
competitive inhibition
feedback inhibition
inborn errors of metabolism
digestive enzymes
brush border enzymes
hormone
target cell
second messenger

Study Questions

1. Although most enzymes are proteins, it is now known that some enzymes are composed of _______.
2. In general, most digestive enzymes catalyze what type of chemical reaction?
3. The specific term for ligands that bind to the active sites of enzyme molecules is _______.
4. Amylases are enzymes involved in the breakdown of what food molecule?
5. How does the enzyme speed up a chemical reaction?
6. Any substance that accelerates a chemical reaction without itself undergoing any permanent chemical change is defined as a _______.
7. Different species of the same enzyme are _______.
8. Inactive forms of an enzyme are termed _______.
9. In the small intestine, trypsinogen is converted to trypsin by _______.
10. Many of the coenzymes in cells are derivatives of vitamin _______.
11. An enzyme that has more than one binding site is termed an _______.
12. A digestive juice that is strongly acidic (pH 1–2) is _______.

Name____Date____________________

CHAPTER 9

Key Terms

intracellular fluid
extracellular fluid
interstitial fluid
solution
solvent
solute
true solution
crystalloid
colloidal solution
Tyndall effect
suspension
compartmentation
vesicle
integral protein
peripheral protein
glycocalyx
permeability
diffusion
equilibrium
concentration gradient
semipermeable membrane
osmosis
osmotic pressure
colloid osmotic pressure
isotonic
hypotonic
hypertonic
hemolysis
crenation
edema
dialysis
carrier-mediated transport
facilitated diffusion
active transport
uniport
cotransport
symport
antiport
ion pumps
Na^+-linked active transport
endocytosis
exocytosis
phagocytosis
pinocytosis
receptor-mediated endocytosis
coated pits

Study Questions

1. The solutes of true solutions are _______.
2. Protein molecules embedded in the lipid bilayer of the cell membrane are called _______.
3. Oligosaccharide groups on membrane proteins and lipids always face the _______ surface of the cell membrane.
4. The degree to which a membrane allows substances to move across it is described by the term _______.
5. Red blood cells suspended in a test solution are observed to lose water and shrink; this effect is called _______.
6. From the effect observed in Question 5, it may be concluded that the test solution was _______ to the cells.
7. What type of solutes will pass through a dialyzing membrane?
8. What feature is common to both facilitated diffusion and active transport?
9. How do facilitated diffusion and active transport differ?

10. What is the source of energy for secondary active transport?
11. Exocytosis is principally linked with the cellular function of ______.
12. Clathrin-coated vesicles are seen inside a cell. What process has taken place?
13. An example of an antiport system that is present in most cells in the body is the ______.

Name________________________________ Date____________________

CHAPTER 10

Key Terms

metabolism
catabolism
anabolism
exergonic reactions
endergonic reactions
free energy
Calorie
substrate-level phosphorylation
oxidative phosphorylation
high-energy compounds
electron carriers
glycolysis
citric acid cycle
electron transport chain
oxygen debt
chemiosmotic theory
proton gradient
respiratory quotient

Study Questions

1. The term *metabolism* is defined as ______.
2. The series of reactions in which glucose is converted to pyruvate is called ______.
3. Do the reactions referred to in Question 2 require oxygen?
4. All of the reactions of the citric acid cycle take place in what part of the cell?
5. In photosynthesis, plants directly utilize ______ to synthesize energy-rich food molecules.
6. The citric acid cycle is initiated when a molecule of ______ condenses with oxaloacetic acid.
7. Name *two* important electron-carrying coenzymes that are derivatives of B complex vitamins.
8. During strenuous muscular activity, the NADH formed during glycolytic reactions must be reoxidized to NAD^+ if glycolysis is to continue. This reoxidation involves the conversion of ______ to ______.
9. Iron-containing heme groups are found in which enzymes of the electron transport chain?
10. As electrons flow through the electron transport chain, protons are pumped from the ______ of the mitochondrion to the ______.
11. Which unit of the F_1/F_0 complex is the proton channel? Which is the ATP synthase?
12. In glycolysis, there is a net gain of ______ ATP per molecule of glucose. What type of phosphorylation is involved in this synthesis of ATP?
13. The final electron acceptor of the electron transport chain is ______.
14. The oxidation of $FADH_2$ by the electron transport chain yields ______ moles of ATP per electron pair.

Name____ Date____________________

CHAPTERS 11 and 12

Key Terms

nucleic acids	ribosomal RNA	translation	intron
nucleotide	messenger RNA	codon	exon
nitrogenous base	transfer RNA	genetic code	operon
DNA/RNA polymerase	template	anticodon	
double helix	replication	genome	
complementary bases	transcription	gene	

Study Questions

1. The pentose sugar present in DNA is ______.
2. The addition of a pentose sugar to a pyrimidine or purine produces a ______.
3. Besides being the building blocks of nucleic acids, nucleotides function in the cell as ______ and ______.
4. The chemical bonds between complementary bases of a DNA molecule are ______.
5. In RNA, the pyrimidine ______ is complementary to adenine.
6. Three major species of RNA are ______, ______, and ______.
7. Most genes in higher organisms have both introns and exons. The ______ are the non-coding sequences.
8. Transcription is defined as ______.
9. The function of the operon is controlled by a ______ gene.
10. Translation takes place on which organelles of the cell?
11. The function of the promoter site in the operon is ______.
12. If a given sequence of bases in a DNA molecule is CGATTGA, then the complementary sequence in an RNA molecule would be ______.
13. During protein synthesis, amino acids are delivered to the translation machinery by ______.
14. Since more than one codon may specify the same amino acid, the genetic code is said to be ______.
15. The genetic code codes for a set of ______ amino acids.
16. The exposed base triplet on a tRNA molecule is called the ______.
17. Which species of RNA carries the 'recipe' (transcribed from DNA) for a given protein?

Name____ Date________________________

CHAPTER 13

Key Terms

allele
mutation
mutagen
inborn errors of metabolism
bacteriophages
oncogenic viruses
oncogenes
recombinant DNA
plasmids
clone
gene therapy
prokaryote
eukaryote

Study Questions

1. A genetic disease that is due to a mutation in a hemoglobin gene is ______.
2. DNA viruses that infect bacteria are called ______.
3. Genes that may promote malignant transformation of cells are called ______.
4. Convenient vehicles for producing recombinant DNA are small bacterial structures called ______.
5. The genetic disease, phenylketonuria (PKU), is due to ______.
6. The human immunodeficiency virus (HIV) is a good example of an RNA virus that can transcribe DNA. Viruses of this type are termed ______.
7. The process by which RNA synthesizes DNA is termed ______.